Weather Forecasting as a Problem in Physics

The MIT Press
Cambridge, Massachusetts,
and London, England

Weather Forecasting as a Problem in Physics

Andrei S. Monin

Translated by
Paul Superak

Originally published by "Nauka" Publishing Company, Moscow, in 1969 under the title "Prognoz pogody kak zadacha fiziki"

English translation
Copyright © 1972 by
The Massachusetts Institute of Technology

This book was designed by the MIT Press Design Department.
It was set in Monophoto Baskerville 169/312,
printed on P&S Offset
and bound in Interlaken ALI-555 Matte
by Halliday Lithograph Corp.
in the United States of America.

Library of Congress Cataloging in Publication Data

Monin, Andreĭ Sergeevich.
    Weather forecasting as a problem in physics.

    Translation of Prognoz pogody kak zadacha fiziki.
    Bibliography: p.
    1. Weather forecasting.   I. Title.
QC995.M7713   551.6'3   72-6129
ISBN 0-262-13083-1

Contents

Foreword to the
English Translation

This book, which appeared in its original Russian three years ago, has not lost its essential timeliness. Professor A. S. Monin is perhaps best known to the West for his original and striking contributions to the theory of turbulence. However, here he has undertaken a systematic and rather unique accounting of the evolution of the dynamic and physical basis for large-scale atmospheric modeling appropriate for the simulation of atmospheric motions spanning the spectrum from short-period to climatic time scales. Of necessity, details are sacrificed in order to preserve the broad sweep, but no compromise has been made in literature citations, which are copious and quite complete. The few exceptions have to do with Monin's hitherto unpublished ideas, which he develops here *in extenso*.

Monin liberally intersperses his exposition with strongly expressed opinions that will find sympathy or dissent from his readers quite independently of national boundaries. Whether or not one agrees, his views will be stimulating. This monograph reveals quite clearly that we in the West are not nearly as aware of the Russian literature on the subject as they are of ours. Monin points out precedents set by Soviet authors that are not generally known to the solely English-reading community of scientists. Not least among these is Monin himself, who has contributed to specialties ranging wide of his acknowledged expertise in the boundary layer. It will not be surprising if this volume creates a secondary demand for translations of many of the original Russian papers.

We have here an important addition to the literature, filling a glaring void in the English as well as the Russian. It will be particularly useful to beginning graduate students in providing a comprehensive assessment of the large-scale problem and the solutions that have thus far been forthcoming. Furthermore, it eloquently demonstrates the inevitable necessity for interactions among the contributing disciplines.

Joseph Smagorinsky
31 August 1971

Preface

Several years ago the author became acquainted with an article by the famous Soviet theoretical seismologist V. I. Kaylis-Borok. In this article, "Seismology and Logic," Kaylis-Borok presented a program for the complete cybernetization of seismology. A comparison with the state of atmospheric physics led the author to conclude that it would be highly useful to write a book called "Meteorology and Logic" as well. For the present it would not be a plan for the cybernetization of meteorology—such a task would be several orders of magnitude more voluminous than for seismology—but a plan for the logical formulation of the main problems of contemporary meteorology.

Every thinking scientific worker (it is desirable that these two adjectives turn out to be synonyms more often) should be able, first of all, to pose questions, at least to himself—not only the questions How? and Why?, but the more difficult question, Now what? and the most difficult question, What here is most important? It seems that the sum total of such fundamental questions, arranged in some natural order, is the logic of a given science.

In the author's opinion the primary task of contemporary meteorology is the long-range prediction of weather. It appears that the main questions in this area (e.g., What periods of weather forecasting are long? or Why do long-term weather changes take place?) were first being posed only in the postwar years, and even then not all of them were asked right away. Two decades of reflection on these themes, lectures and publications on several issues, discussions with friends and colleagues (A. M. Obukhov, A. M. Yaglom, E. M. Feygel'son, L. A. Diky, and B. L. Dzerdzeyevsky in the Academy of Sciences of the U.S.S.R., E. N. Blinova and I. A. Kibel' in the Hydrometeorological Service), and conversations with Jule Charney, Norman Phillips, Joseph Smagorinsky, Philip Thompson, and Yale Mintz led the author to attempt to write "Meteorology and Logic." The result is this book. The persons listed, as well as the author, having devoted their labor to the introduction of physical and mathematical thinking into meteorology, often find themselves in a thankless situation: The weather services regard them jealously, and mathematicians and physicists treat them condescendingly. Addressing the present book to both groups, the author would be happy if it should prove capable of shaking both these attitudes.

Some of the positions taken in this book are proven theorems (e.g., the theorem of potential vorticity conservation), while others represent only the opinion of the author and are subject to dispute (e.g., the uselessness of returning to the primitive equations, the existence of free cumulus convection, the crucial significance of the numerical experiment method, the importance of the predictability problem, the lack of influence by solar activity on the weather, and so on). In the belief that scientific debates are conducted too rarely, since they can in the final analysis help in ascertaining the truth, the author decided not to be afraid of expressing opinions on debatable issues.

Quite a few literature sources are cited in the book but without any pretense of historical methodology, assignment of priority, completeness, or choice of the most important works.

The author considers it his pleasant duty to express his thanks to L. A. Diky for his valuable help, to B. L. Gavrilin, who read several sections in the manuscript, to G. S. Golitsyn, who supplied material for the section on the atmospheres of other planets, and to A. N. Gezentsvey, L. M. Belova, and N. I. Solntseva for their aid in putting the manuscript in order.

A. Monin

1 **Introduction**

1. A Short History of the Problem

There is no intent here to set forth a detailed history of scientific methods of weather forecasting (to say nothing of such unscientific methods as those exemplified in the *Old Farmer's Almanac*). In this account only the five events that seem to have played the most important role in the formulation of modern weather-forecasting theory will be cited. (For more detailed historical information see, for example, Khrgian's book.[1])

The first of these events was the organization of a weather service by Urbain Leverrier, director of the Paris Astronomical Observatory. Leverrier received a contract from the French government after the catastrophic storm of 14 November 1854 in Balaklava. On 19 February 1855, he became the first person to compile a weather map for the same day.

The second event was the publication of an article by Vilhelm Bjerknes, entitled "The Problem of Weather Forecasting, Considered from the Point of View of Mathematics and Mechanics" (1904). In his article weather prediction was formulated for the first time as an initial-value problem for the hydrodynamic equations of a baroclinic fluid.* In the works of Jakob Bjerknes in 1917–1919 and subsequent work by the so-called Bergen school, two important concepts were developed: that of atmospheric fronts—i.e., the interfaces between different air masses—and that of cyclone formation resulting from wave instability on frontal interfaces. These works became the basis for modern synoptic methods of short-range weather prediction. However, these concepts have up to now had little significance in numerical forecasting, though for purely technical reasons. The

---

*A fluid is called barotropic if its density ρ is a function only of the pressure p; otherwise, it is called baroclinic. A real atmosphere is baroclinic; its density ρ depends on the temperature T as well as on p. If the humidity is ignored, then Clapeyron's equation holds: $p = \rho R T$ ($R \approx 0.287 \, \mathrm{J/g °C}$ is the gas constant for dry air). For a nonviscous fluid in a field of potential forces, V. Bjerknes's theorem holds: $d\Gamma/dt = -\oint_L (dp/p)$ (where $\Gamma = \oint \mathbf{V} \cdot d\mathbf{S}$ is the circulation integral of the velocity vector along the closed contour L). From this it follows that in a baroclinic fluid the intersection of the surfaces $p = \mathrm{const}$ and $\rho = \mathrm{const}$ (or $p = \mathrm{const}$ and $T = \mathrm{const}$) leads to vorticity production.

spatial grids used in numerical forecasts have horizontal increments of several hundred kilometers; thus they do not allow consideration of narrow zones with large hydrodynamic field gradients—in particular, atmospheric fronts. In order to take such zones into account in numerical forecasts, it is necessary to develop other methods.

The third major event was the publication of Lewis Richardson's *Weather Prediction by Numerical Process* (1922). In this book Richardson presented the first attempt at forecasting weather by using numerical solutions of the hydrodynamic equations (in the same way that astronomers predict the positions of the planets by solving equations for the dynamics of the planetary system). His attempt was unsuccessful: Completed after very long calculations, his prediction for one day (20 May 1910, for the Nuremberg-Augsburg area) was unsatisfactory. From today's vantage point the reasons for this failure are clear: the incompleteness of his initial data (at that time only ground data were available from a scanty network of stations in Europe); the imperfection of finite-difference schemes (for example, he was not aware of the Courant-Friedrichs-Lewy criterion, established in 1928, for correlating space and time computational intervals); the needless physical complexity of the equations he integrated, which described not only the motions important for the weather (the so-called synoptic-scale processes) but in addition all sorts of "noise," such as acoustic waves, plant respiration, and so on.

This latter difficulty was overcome in 1940 by the work of Kibel',[2] the publication of which constitutes the fourth milestone in this brief historical account. Kibel' proposed a basic principle for simplifying the hydrodynamic equations: the asymptotic "quasi-geostrophic expansion." This analysis makes it possible to filter out from solutions of the equations the meteorological noise that is unessential to weather forecasting. The creation of a hydrodynamic theory of short-range weather forecasting was subsequently based on this very principle.

In 1940, trying to limit his calculations to only the lower portion of the atmosphere (the troposphere), Kibel' imposed at the upper boundary of this layer (the tropopause) an artificial condition that does not follow from the laws of hydrodynamics. The equations of the "quasi-geostrophic approximation," free from this limitation, were derived only in the postwar years. They appeared (marking the fifth event in this account) in papers published almost simultaneously by Obukhov[3] and Charney.[4, 5]

These equations were immediately employed in practical calculations; weather prediction was one of the problems John von Neumann had in mind when high-speed calculating machines were being devised; he was coauthor of one of the first papers[6] on the integration of the quasi-geostrophic equations with an electronic computer. The most outstanding achievement of atmospheric physics in the postwar years has been the formulation of a hydrodynamic theory for the short-range prediction of meteorological fields. By now, this theory has been published in a number of books[7-10] especially devoted to numerical methods of weather prediction.

As for the physical basis of long-range weather prediction, the history of its development is much shorter; it dates from a 1943 paper by Blinova.[11] So far no definitive formulations in this area have been reached. One fundamental difference between short- and long-range forecasts (which also accounts for the differences among the various theories) is, in our opinion, that the adiabatic approximation is sufficient for the hydrodynamic equations in short-range forecasts, while nonadiabatic processes constitute the very essence of long-range weather changes.[12]

2. Scales of Weather Processes

We now introduce several numerical characteristics that will later be useful in applying the hydrodynamic equations to the atmosphere. The mass M of the atmosphere is 5.3×10^{21} g. The total kinetic energy of its motion E has a magnitude on the order of 10^{21} J (roughly speaking, the energy of an individual cyclone is smaller by two orders of magnitude; by way of comparison, one megaton of TNT $\approx 4 \times 10^{15}$ J). According to empirical estimates by Borisenkov,[13] for the Northern Hemisphere $E = 4 \times 10^{20}$ J in the winter and 1.9×10^{20} J in the summer; for the Southern Hemisphere, the magnitudes are 7.1×10^{20} J (winter) and 3.9×10^{20} J (summer). Similar estimates were made by Pisharoty[14] and Gruza[15] (Gruza demonstrated that in the troposphere on the average more than 70% of the kinetic energy is due to zonal, i.e., latitudinal, motion, and less than 30% is due to meridional motion; approximately half of the energy is associated with the mean zonal circulation and half with deviations from it). The kinetic energy per unit of mass, E/M, is of the order of 10^2 J/kg $= (10$ m/sec$)^2$; therefore $U = 10$ m/sec is taken as the typical speed of air motion in synoptic-scale processes.

Solar heat is the primary energy source for the atmospheric processes. The power that the sun transmits to earth is equal to 1.8×10^{14} kW, but about 40% of the sun's radiation is immediately reflected back into space. Thus, as an initial figure one should take 1×10^{14} kW, or for the average power per unit surface area, 20 mW/cm^2. Only a small fraction of this energy is converted into the kinetic energy of atmospheric motion. According to empirical estimates by Palmén,[16,17] the rate $\partial E/\partial t$ at which potential energy is converted into kinetic energy is approximately equal to 2×10^{12} kW for the atmosphere as a whole. So the efficiency coefficient of the "atmospheric machine" amounts to only about 2% (in individual cyclones $\partial E/\partial t \sim 1 \times 10^{11}$ kW to 2×10^{11} kW, outside of which there is on the average a slow reverse conversion of kinetic into potential energy). According to these figures, the average rate of kinetic energy generation in a unit of mass, $(1/M)\,(\partial E/\partial t)$, is equal to 4 cm^2/sec^3. The average specific rate ε of kinetic energy dissipation into heat (due to friction) must be of the same order of magnitude; and, in fact, by an independent method Brunt[18] in 1926 obtained $\varepsilon \sim 5$ cm^2/sec^3 for the troposphere.

The typical energy conversion time is $\tau = \left((1/E)\,(\partial E/\partial t)\right)^{-1} = (10^{21}$ J$)/(2 \times 10^{12}$ kW$) = 5 \times 10^5$ sec, i.e., approximately a week. In synoptic processes the typical relaxation time due to viscosity is of the same order. Indeed, consider the following situation: In the

interval of scales L a cascade process takes place with energy transmitted from large-scale to small-scale motions. It proceeds at a
constant rate ε (which is independent of L). The effective "viscosity"
under these conditions will have the form $v(L) \sim \varepsilon^{1/3} L^{4/3}$. This is
Richardson's[19] so-called four-thirds law, which is valid for nearly
the entire spectrum of scales of atmospheric motion, from millimeters to thousands of kilometers (Fig. 1). As a result of this
"viscosity," the relaxation time will be $\tau(L) \sim L^2/v(L) \sim \varepsilon^{-1/3} L^{2/3}$.
According to Obukhov,[3] the typical length scale for the synoptic
processes is on the order of $L_0 = c/l$, where c is the sound speed and
$l = 2\omega \cos \theta$ is the so-called Coriolis parameter ($\omega = 7.29 \times
10^{-5}$ sec^{-1} is the angular velocity of the earth's rotation, and θ is
the latitude plus $\pi/2$); in the temperate latitudes $L_0 \sim 3000$ km.
Using this value of L_0 and the value of ε given above, we get for the
synpotic processes $\tau(L_0) \sim \varepsilon^{-1/2} L_0^{2/3} \sim 3 \times 10^5$ sec. Note that the
Eulerian time scale for synoptic processes $\tau_1 = L_0/U$ is of the same
order of magnitude (since, in the west-east flow in the temperature
zones the middle atmospheric layers complete a full circuit around
the earth in a few weeks while the Eulerian atmospheric time scale
is on the order of one month).

It will also be useful to cite some data on the role played by the
humidity of the air in the energy reserve of the atmosphere.
According to Rudloff[20] (and Neik[21] gives similar figures), the
atmosphere contains an average of 1.24×10^{19} g of moisture, equiv-
lent to a 24-mm layer of precipitated water. (The ocean contains
1.37×10^{24} g of water. Glaciers add another 2.9×10^{22} g of ice;
their melting would raise the level of the ocean by 80 m.) The average
yearly amount of precipitation is 3.96×10^{20} g (2.97×10^{20} g on
the oceans and 0.99×10^{20} g on dry land). This is equivalent to a
780-mm layer of water. (Thus the water vapor in the atmosphere is
replaced on an average of $780/24 \approx 32$ times per year, or once every
11 days.) The annual amount of evaporation is the same; only the
ocean's share is 3.34×10^{20} g, and the evaporation from dry land
amounts to 0.62×10^{20} g. Runoff from the land accounts for

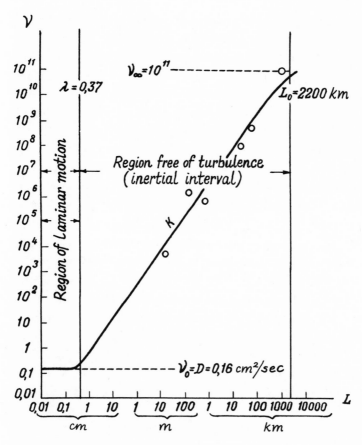

Fig. 1   The vertical diffusion coefficient $v(L)$ as a function of the turbulence scale L. Empirical points after Richardson.[19]

0.37 × 10^{20} g. If we take for the latent heat of vaporization the value 2.4 × 10^3 J/g, then the power of heat expended in vaporization amounts to 3 × 10^{13} kW—that is, 30% of the 10^{14} kW of solar heat absorbed by the earth. This turns out to be 15 times greater than the rate of kinetic energy generation in the atmosphere (2 × 10^{12} kW).

3. The Spectrum of Atmospheric Processes

Meteorological elements—the velocity of air motion, temperature, pressure, humidity, and so on—fluctuate with time. Their oscillations have components with periods ranging from a fraction of a second to tens of thousands of years. The entire spectrum of these oscillation periods can be divided into the following nine intervals.

3.1 *Micrometeorological oscillations,* with periods ranging from a fraction of a second to one minute. The greatest contributor to this category is small-scale turbulence. In the surface layer of the atmosphere its energy spectrum $fS(f)$ has a maximum value when the period τ is of the order of one minute (f is the frequency, τ the period of oscillation, and $S(f)$ the spectral energy density). This corresponds to a scale of the horizontal-turbulent inhomogeneities $L = U\tau_{\max} \sim 600$ m. For $f \gg 1/\tau_{\max}$, the wind-velocity spectra satisfy the "five-thirds law" of Kolmogorov and Obukhov,[22] $S(f) \sim (\varepsilon^{2/3}/U)(f/U)^{-5/3}$. The temperature spectra have a similar form,[23] $S_T(f) \sim (N\varepsilon^{-1/3}/U)(f/U)^{-5/3}$, where $N = \chi\overline{(\nabla T)^2}$ is the rate of decay of temperature inhomogeneities, T the temperature, and χ the molecular conductivity. In the region of the maximum frequency of turbulent fluctuations $f \sim U\varepsilon^{1/4}\nu^{-3/4}$ (ν is the molecular viscosity), the turbulence spectrum drops sharply.

In addition to turbulence, the category of micrometeorological oscillations includes acoustic and short-period gravitational waves (with relatively small amplitudes). According to the theory (see below), gravitational waves predominantly have periods longer than 330 sec, and acoustic waves have periods shorter than 300 sec. This explains the minimum near $\tau \sim 300$ sec in the spectrum $fS_p(f)$ of the pressure fluctuations. Golitsyn[24] has plotted this spectrum according to microbarogram data (see Fig. 2, where the scale unit

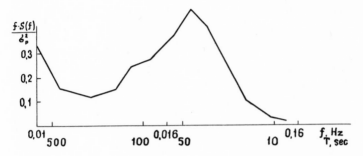

Fig. 2   Spectrum of pressure micropulsations. After Golitsyn.[24]

on the vertical axis is the variance of the pressure microfluctuations $\sigma_p{}^2$, where $\sigma_p \sim 10^{-2}$ mb).

3.2. *Mesometeorological oscillations*, with periods ranging from a minute to an hour. Here intense oscillations of the meteorological elements (including, for example, their oscillations during thunderstorms or gravitational waves with large amplitudes) are relatively rare; therefore, there is usually a pronounced broad minimum in the spectrum $fS(f)$ within this interval. For a summary of data concerning this interval, see the paper by Kolesnikova and Monin.[25] The mesometeorological minimum is clearly defined, for example, in the spectrum given in Fig. 3. Van der Hoven[26] constructed this spectrum of horizontal wind velocity with data from measurements taken on the 125-meter Brookhaven weather tower. The minimum in this spectrum corresponds to a period τ on the order of 20 min and to a scale $L = U\tau$ on the order of the effective thickness of the atmosphere $H \sim 10$ km (the lowest 10 km layer of the atmosphere contains 80% of the atmosphere's mass). The minimum separates the quasi-two-dimensional (quasi-horizontal) synoptic inhomogeneities with scales $L \gg H$ from the essentially three-dimensional (quasi-isotropic) micrometeorological inhomogeneities with scales $L < H$. The presence of this minimum allows one to obtain relatively stable mean values for wind velocity, temperature, and so on, in micrometeorology. This is done by averaging over the range of

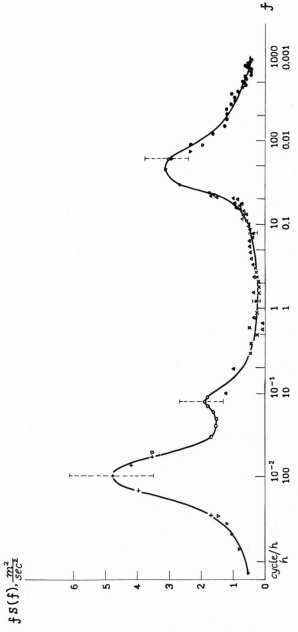

Fig. 3 Spectrum of the horizontal wind velocity. After Van der Hoven.[26] Some experimental points are shown on the graph; see reference 26.

periods lying within the mesometeorological interval (in practice, τ is taken to be 10 to 20 min).

3.3. *Synoptic oscillations*, with periods ranging from many hours to several days and with an energy-spectrum maximum near $\tau = 4$ days (see Fig. 3). (Van der Hoven considers the relative maximum at $\tau = 12$ h insignificant.) Diurnal fluctuations also fall into this interval. These are seen, for example, in the temperature spectrum in the form of a diurnal spectral line and in the pressure spectrum in the form of diurnal and semidiurnal lines. In the high-frequency part of the synoptic interval there is a cascade energy transfer along the spectrum from large-scale to small-scale motions. This transfer is due to the hydrodynamic instability of the quasi-horizontal synoptic motions with large Reynolds numbers $\mathrm{Re} = UL/\nu$. (On the low-frequency end of the synoptic interval, there apparently is an energy transmission in the opposite direction—from synoptic motions to motions of a still larger scale, that is, to the motions of the general atmospheric circulation.[27,28])* Moreover, all motions on the synoptic scale generate microturbulence directly (that is, by-passing all the motions of the intermediate scales) and continuously. This generation results from the hydrodynamic instability of the vertical inhomogeneities in the wind field, especially near the surface

---

*In the case of two-dimensional turbulence in an incompressible fluid there are two invariants of motion—energy E and vorticity ω. In an equilibrium range of spectrum two parameters are essential—the rate of energy transfer through the spectrum range $\varepsilon \sim d\overline{U^2}/dt$ and the rate of mean square vorticity transfer $\varepsilon_1 \sim d\overline{\omega^2}/dt$, and the length scale $L = (\varepsilon/\varepsilon_1)^{1/2}$ can be defined. Then the equilibrium energy spectrum is of the form $E(k) = C(kL)\varepsilon^{2/3}k^{-5/3}$. There is a possibility that in the high-frequency part of the equilibrium range only ε_1 is essential, i.e., $C(kL) \sim (kL)^{-4/3}$ and $E(k) \sim \varepsilon_1^{2/3}k^{-3}$. Some confirmation of this spectral law was found in numerical experiments on two-dimensional turbulence by G. K. Batchelor (*Proc. Int. Symp. High-Speed Computation in Fluid Dynamics*, 1969) and D. K. Lilly (NCAR manuscript No. 68–234, Nov. 1968); see also R. Kraichnan [*Phys. Fluids* **10**, 1417 (1967)] and F. Baer [*Proc. Int. Symp. Numerical Weather Prediction, Tokyo, 1969*, pp. 95–100 (1)]. In the high-frequency part of the synoptic range, atmospheric motions are quasi-two-dimensional, and the law $E(k) \sim k^{-3}$ may be applicable. This possibility is at least qualitatively confirmed by J. Smagorinsky's calculations [*Monthly Weather Rev.* **91**, 99–164 (1963)].

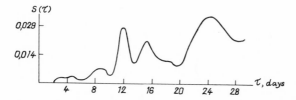

Fig. 4   The spectral density of oscillations of the circulation index. After Monin.[30]

of the earth and in the so-called jet stream, where the vertical gradients of the wind velocity have the greatest magnitude.

By regarding microturbulence as a dissipative factor for the synoptic motions, it can be characterized by an effective viscosity coefficient v_{turb}. The minimum scale of synoptic motions capable of overcoming this viscosity is equal to $L_{min} \sim \varepsilon^{-1/4} v_{turb}^{3/4}$. The presence of the mesometeorological minimum indicates[25] that $L_{min} \gtrsim H$.

3.4. *Global oscillations,* with periods ranging from weeks to months. These are of the greatest interest for the problems of long-range weather prediction, but they still have been studied very little. So far, perhaps only the index cycle has been more or less precisely determined. This is the oscillation cycle of the planetary circulation between states of intense zonal flow (the west-east transfer) with weak meridional mixing and states of weaker zonal flow with intense meridional mixing. This cycle was traced, for example,[29-31] through the oscillations of the circulation index $\alpha = u/a \sin \theta$, i.e., the mean angular velocity of atmospheric rotation at temperate latitudes with respect to the earth. (Here u is the mean zonal velocity along a line of latitude and a is the radius of the earth.) The period of the index cycle is close to two weeks. Figure 4 illustrates the spectral density $S(\tau)$[30, 31] of the variations in the circulation index. A sharp maximum occurs at $\tau = 12$ days. (Note that oscillations of the circulation index are, strictly speaking, not stationary but are a periodic random process[31] with a period of one year.)

3.5. *Seasonal oscillations* are the oscillations with a one-year period and its harmonics.

3.6. *Interannual oscillations*, with periods on the order of several years. The spectrum of these variations is still virtually unstudied. (Consider the 26-month rhythm of oscillations in the equatorial stratosphere, observed by several authors, and also the hypothesis that the 11-year cycle of solar activity is manifested in the earth's weather, which in my opinion lacks convincing evidence.) According to Kolesnikova and Monin,[32] the amplitude of interannual fluctuations in the average annual values of temperature and several other meteorological elements usually is 15% to 30% of the amplitude of their seasonal and irregular variations within the year.

One should not confuse interannual variations with climatic variations. If all the oscillations (3.1 through 3.6) discussed above are called short-period oscillations, then climate is the statistical regime of short-period meteorological oscillations, the spectrum of which can be divided into the following three intervals.

3.7. *Intrasecular oscillations*, a clear example of which is the warming trend (which is now apparently ending) in the first half of the twentieth century. Here a connection is observed between climatic changes and the character of the general circulation of the atmosphere; according to Dzerdzeyevsky,[33] zonal types of circulation in the Northern Hemisphere were observed less often and meridional types more often in 1900–1930 than in 1930–1950. Explaining the source of climatic warming in the twentieth century is a timely problem for the physical theory of the climate.

3.8. *Intersecular oscillations*. These include[34] the warming after the end of the Ice Age (65 centuries B.C.). This warming led to the so-called climatic optimum of the fortieth to twentieth centuries B.C., with the subsequent worsening in the Sub-Atlantic period (tenth century B.C. to the third century A.D.). The climate improved during the fourth to tenth centuries A.D.,* deteriorated again in the thirteenth and fourteenth centuries, improved in the fifteenth to sixteenth centuries, and deteriorated in the seventeenth to nineteenth centuries (the so-called Little Ice Age).

3.9. *The glacial periods of the Pleistocene*: Günz (500–475 millennia B.C.), Mindel (425–325 millennia B.C.), Riss (200–125 millennia B.C.) and Würm (60–29 millennia B.C.). During these periods the average temperature of the lowest air layer (today $+15°C$) apparently dropped by $10°$. So many causes have been proposed for the glaciation of the earth (including, for example, Simpson's hypothesis[35] that the solar radiation increased, causing more evaporation, cloudiness, and snowfall), that perhaps one should not try to explain the glaciation phenomena but rather the absence of glaciers during 90% of the post-Cambrian time.

*The small climatic optimum of the eighth to tenth centuries coincided with the Vikings' colonization of Iceland and Greenland. According to Viking chronicles, there was little glaciation in the Arctic. This testifies against the hypothesis that if the Arctic melts, it will no longer freeze; this hypothesis comes from overestimating the role of the local albedo (the reflective property of the underlying surface) in the Arctic environment. Considering that the Arctic (bounded by the 72d parallel) comprises only 2.5% of the earth's surface area, it is more natural that, on the contrary, local conditions in the Arctic (including its glaciation and consequently its albedo) are not the cause but the result (an indicator) of the state of the general circulation of the atmosphere.

**The Hydrodynamic
Theory of Short-Range
Weather Prediction**

4. Adiabatic Invariants

In section 2 it was shown that the typical generation time for the kinetic energy of the synoptic processes

$$\tau = \left(\frac{1}{E}\frac{\partial E}{\partial t}\right)^{-1}$$

is on the order of one week. The typical time for the dissipation of the kinetic energy of the synoptic processes $\tau(L_0) \sim \varepsilon^{-1/3}L_0^{2/3}$ is on the same order of magnitude. It is natural to call periods of time $t - t_0 < \tau$ *short* and periods $t - t_0 > \tau$ *long*.[1] Thus, in the theory of short-range forecasts, by their very definition, we may ignore energy sources and sinks; that is, we may use the adiabatic approximation. On the other hand, this approximation is obviously unacceptable in long-range forecasts.

For adiabatic processes— the only type considered in this part of the book—two laws of conservation hold: During the motion of any volume of air V there is conservation of the entropy $\int s\rho\,dV$ (where s is the entropy per unit of mass and ρ the density) and of the "vortex charge" $\int (\mathbf{\Omega}_a \cdot \nabla s)\,dV$ (where $\mathbf{\Omega}_a$ is the absolute vorticity); that is,

$$\frac{ds}{dt} = 0, \qquad \frac{d}{dt}\left(\frac{\mathbf{\Omega}_a \cdot \nabla s}{\rho}\right) = 0. \tag{4.1}$$

The specific vorticity $\Omega = (\mathbf{\Omega}_a \cdot \nabla s)/\rho$ is called the potential vorticity (Rossby[2]). The conservation law $d\Omega/dt = 0$ was first derived by Ertel;[3] see also Charney,[4] Arakawa,[5] and Obukhov.[6,7]

Hollmann[8] pointed out three other independent combinations of hydrodynamic fields that are conserved in adiabatic processes:

$$\psi_3 = \left(\frac{\nabla s \times \nabla \Omega}{\rho} \cdot \mathbf{w} \right),$$

$$\psi_4 = \left(\frac{\nabla s \times \nabla \psi_3}{\rho} \cdot \mathbf{w} \right), \qquad (4.2)$$

$$\psi_5 = \left(\frac{\nabla s \times \nabla \Omega}{\rho} \cdot \nabla \psi_3 \right),$$

where $\mathbf{w} = \mathbf{v}_a - \nabla W$, with $\mathbf{v}_a$ the absolute velocity and $W = \int_0^t \Lambda \, dt$ the so-called action (with Lagrangian $\Lambda = v_a^2/2 - \varphi - \eta$, where φ is the gravitational potential and η the specific enthalpy).

For dry air, the entropy increment ds is given by the formula

$$ds = c_p d \ln \left(\frac{p^{1/\kappa}}{\rho} \right) = c_p d \ln \theta, \qquad \theta = T \left(\frac{p_0}{p} \right)^{(\kappa-1)/\kappa}, \qquad (4.3)$$

where p and T are the pressure and temperature, respectively, $\kappa = c_p/c_v$ is the ratio of the specific heat capacities at constant pressure and at constant volume ($c_p/c_v = 1.41$, $c_p = 1.003$ J/g°C, and $c_v = 0.717$ J/g°C); θ is the so-called potential temperature; p_0 is the standard pressure (usually $p_0 = 1000$ mb $= 10^6$ dyn/cm^2). The moisture content of the air is always small; therefore, in many (but not all) calculations the entropy of unsaturated moist air can be determined by the same formula (4.3).

The partial pressure of the saturated water vapor $e_m(T)$ is determined by the Clausius-Clapeyron formula

$$\frac{1}{e_m} \frac{de_m}{dT} = \frac{\mathscr{L}}{R_v T^2}, \qquad \mathscr{L} = \mathscr{L}_0 - (c_w - c_{pv})(T - T_0) \qquad (4.4)$$

(where $\mathscr{L}$ is the heat of vaporization; $c_{pv} = 1.81$ and $R_v = 0.461$ J/g°C are the specific heat at constant pressure and the gas constant of water vapor; $c_w = 4.19$ J/g°C is the specific heat of water); this pressure assumes values ranging from 0.509 to 42.47 mb as the temperature increases from -30 to $+30$°C. The specific humidity q, i.e., the ratio of the densities of the water vapor and the moist air, usually does not exceed 3–4%. The content of liquid water and ice in clouds is usually appreciably smaller than the vapor content.

For saturated air, on the other hand, it is sufficient to add the term $(\mathscr{L}/T)\,dq_m$ to the expression (4.3) for ds, where $q_m \approx (R/R_v)\,(e_m/p)$ is the saturation specific humidity. If, however, we ignore the influx of heat due to phase changes of the moisture, we can often completely disregard the humidity of the air; so far this has been the most frequent practice in operative short-range numerical weather forecasts. Accordingly, we shall for the time being use formula (4.3) for ds.

According to (4.1), any function of s and Ω will be an adiabatic invariant, i.e., a conservative characteristic of moving air particles undergoing adiabatic processes. It is convenient to choose two such independent functions for the Lagrangian coordinates of the air particles. The corresponding coordinate surfaces divide the atmosphere into tubes. Air does not flow through the walls of these tubes, so that the adiabatic evolution of the atmosphere will consist only in the deformation of these tubes. The prediction of these deformations is the basis of short-range weather forecasting.

It is convenient to choose the potential temperature θ as one of the Lagrangian coordinates. Its changes are much greater in the vertical direction than in the horizontal direction, so that the vector $\nabla\theta$ has an approximately vertical orientation. Thus we can put

$$\mathbf{\Omega}_a \cdot \nabla\theta \approx \Omega_{az}\frac{\partial\theta}{\partial z} = (\Omega_z + l)\frac{\partial\theta}{\partial z},$$

where l is the above-mentioned Coriolis parameter, and Ω_z is the vertical component of the relative vorticity. Using the quasi-static approximation, i.e., the equation $\partial p/\partial z = -\rho g$ (g is the acceleration due to gravity; the meaning of this approximation will be explained below), we obtain $\partial\theta/\partial z = (\theta/T)\,(\gamma_a - \gamma)$, where $\gamma_a = [(\kappa - 1)/\kappa]\,(g/R) \approx 10°\text{C/km}$ is the adiabatic temperature gradient, and $\gamma = -\partial T/\partial z$ is the actual temperature gradient. Using these results and following Obukhov,[7] we can choose as the second

Lagrangian coordinate the function

$$\tilde{\Omega} = \frac{\Omega}{c_p R} \frac{p^*(\theta)}{\gamma_a - \gamma^*(\theta)} \approx (\Omega_z + l) \frac{\gamma_a - \gamma}{\gamma_a - \gamma^*(\theta)} \frac{p^*(\theta)}{p}, \tag{4.5}$$

where $p^*(\theta)$ and $\gamma^*(\theta)$ are the standard values of p and γ on the surfaces $\theta = $ const (i.e., the characteristics of the so-called standard atmosphere). As θ changes most rapidly in the vertical direction and $\tilde{\Omega}$ varies most rapidly along a meridian (since usually $|\Omega_z| \ll l$ and $\Omega_z + l \approx 2\omega \cos \theta$), θ and $\tilde{\Omega}$ can replace the vertical coordinate and the latitude. The most understandable graphic representation of the invariant $(\theta, \tilde{\Omega})$-tubes is given by their meridional section. An example of such a section is given in Fig. 5; for the same case, Fig. 6 shows the isolines $\tilde{\Omega} = $ const on one of the surfaces $\theta = $ const.

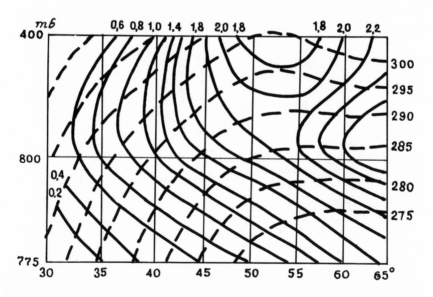

Fig. 5 A meridional section of $(\theta, \tilde{\Omega})$-tubes at longitude $100°\,\mathrm{W}$ for 1 April 1962. After Obukhov.[7] The abscissa is latitude and the ordinate is the atmospheric pressure; $\tilde{\Omega} = $ const is given by the solid lines and $\theta = $ const is given by the dotted lines; $\tilde{\Omega}$ is measured in units of $10^{-4}\,\mathrm{sec}^{-1}$.

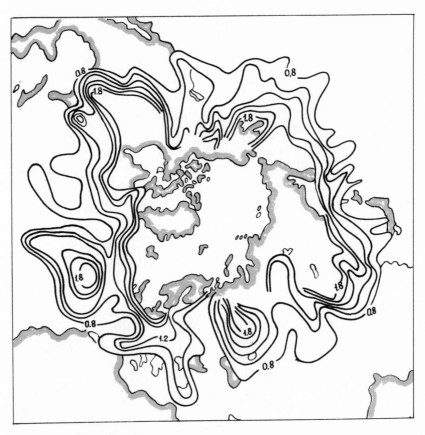

Fig. 6   The configuration of the lines $\tilde{\Omega} = \text{const}$ on the surface $\theta = 300°$ for 1 April 1962. After Obukhov.[7]

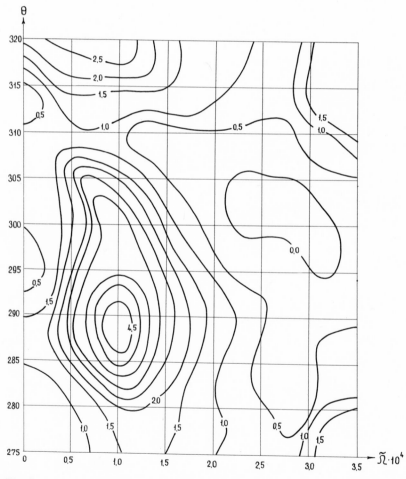

Fig. 7 The probability density $\mu(\theta,\tilde{\Omega})$ from data for the period 1–10 April 1962.

If by $\mu(\theta, \tilde{\Omega}) \, d\theta \, d\tilde{\Omega}$ we denote the fraction of the atmosphere's mass contained in a thin infinite $(\theta, \tilde{\Omega})$-tube, then $\mu(\theta, \tilde{\Omega})$ can be interpreted[8] as the probability density for the values of the Lagrangian coordinates $\theta, \tilde{\Omega}$ of a randomly chosen particle of air. Figure 7 shows an example of the probability distribution $\mu(\theta, \tilde{\Omega})$ from data calculated by Karunin for the period 1–10 April 1962.

In addition to the differential invariants s and Ω (or any two functions of them, for example, θ and $\tilde{\Omega}$), adiabatic processes, as is well known, also have an integral invariant, the *total energy* $\mathscr{E} = \mathscr{K} + \mathscr{N}$, where $\mathscr{K} = \int (\rho v^2/2) \, dV = \int (v^2/2) \, dm$ is the kinetic energy (v is the magnitude of the velocity; dV is the volume element; $dm = \rho \, dV$ is the mass element; and the integral is taken over the entire atmosphere), and $\mathscr{N}$ is the labile energy, i.e., the sum of the potential energy $\mathscr{P} = \int \Phi \, dm$ ($\Phi = gz$ is the potential due to the force of gravity) and the internal energy $\mathscr{I} = \int c_v T \, dm$. In the quasi-static approximation, the potential energy of a vertical column of air with a unit cross section is equal to $\int gz\rho \, dz = -\int z \, dp = \int p \, dz$, so that by the Clapeyron equation $p = \rho R T$, we get $\mathscr{P} = \int R T \, dm$; and since $c_v + R = c_p$, we have

$$\mathscr{N} = \int c_p T \, dm = c_p \int \left(\frac{p}{p_0}\right)^{(\kappa-1)/\kappa} \theta \, dm. \tag{4.6}$$

In adiabatic processes the total mass of the air above any isentropic surface $\theta = $ const does not change, so that the average pressure $p^*(\theta)$ also does not vary on any isentropic surface. For this reason,

$$\mathscr{N}^* = c_p \int \left[\frac{p^*(\theta)}{p_0}\right]^{(\kappa-1)/\kappa} \theta \, dm \tag{4.7}$$

is also constant. The quantity $\mathscr{N}^*$ is that labile energy of the atmosphere that remains if the atmosphere is adiabatically brought to a state with constant pressure along any isentropic surface (and with stable stratification). It is clear that in adiabatic processes not all the labile energy but at most a part of it $\mathscr{A} = \mathscr{N} - \mathscr{N}^*$ that is called the *available potential energy* (Lorenz[9,10]) can be converted into kinetic energy; the magnitude of $\mathscr{A}$ is equal to the weighted average value of the dispersion of θ on the isobaric surfaces. Consequently, the sum $\mathscr{K} + \mathscr{A}$ is an adiabatic invariant.

Suppose the atmosphere is adiabatically brought to a state with indifferent stratification (in which the isentropic surfaces are vertical, i.e., θ doesn't depend on p). The labile energy $\bar{\mathscr{N}}$ in this state can be computed by replacing the factor $p^{(\kappa-1)/\kappa}$ in the integral in (4.6) by its value averaged over the mass of the vertical air column

$$\frac{\int p^{(\kappa-1)/\kappa}\, dm}{\int dm} = \frac{\kappa}{2\kappa - 1}\bar{p}^{(\kappa-1)/\kappa},$$

where $\bar{p}$ is the pressure at the earth's surface; once this is done, we get

$$\bar{\mathscr{N}} = \frac{\kappa}{2\kappa - 1} c_p \int \left(\frac{\bar{p}}{p_0}\right)^{(\kappa-1)/\kappa} \theta\, dm. \tag{4.8}$$

The difference $\mathscr{S} = \bar{\mathscr{N}} - \mathscr{N}$ is called the *macrostability parameter* (Lorenz[11]); it is equal to the weighted average value of the vertical gradient of the potential temperature $-\partial\theta/\partial p$ over the entire thickness of the atmosphere. The magnitude of $\mathscr{S}$ determines the amount of kinetic energy that is released or absorbed in an adiabatic transition from a given stratification to the indifferent one. The difference $\mathscr{K} - \mathscr{S}$ is also an adiabatic invariant.

5. Classification of Atmospheric Motions

In order to construct a hydrodynamic theory of short-range weather forecasting, it is important first of all to ascertain which types of atmospheric motions are possible in adiabatic processes. All these motions have the character of waves, and for their classification it is sufficient to consider the case of small-amplitude waves, i.e., small oscillations of the atmosphere relative to its state of rest (in which the pressure $\bar{p}$, the density $\bar{\rho}$, and the temperature $\bar{T}$ depend only on the altitude z and are related by the static equation $\partial\bar{p}/\partial z = -\bar{\rho}g$ and the Clapeyron equation $\bar{p} = \bar{\rho}R\bar{T}$). In the flat-earth approximation, the equations of motion for small oscillations are of the form

$$\left.\begin{aligned}
\bar{\rho}\frac{\partial u}{\partial t} &= -\frac{\partial p}{\partial x} + l\bar{\rho}v, \\[6pt]
\bar{\rho}\frac{\partial v}{\partial t} &= -\frac{\partial p}{\partial y} - l\bar{\rho}u, \\[6pt]
\bar{\rho}\frac{\partial w}{\partial t} &= -\frac{\partial p}{\partial z} - g\rho,
\end{aligned}\right\} \tag{5.1}$$

where u, v, and w are the components of the velocity vector $\mathbf{v}$, and p and ρ are the perturbations of the pressure and the density (these five functions fully describe small adiabatic oscillations); l, as before, is the Coriolis parameter. The equations expressing continuity and adiabaticity should be added to these equations. Such equations are of the form

$$\frac{\partial \rho}{\partial t} + \operatorname{div} \bar{\rho}\mathbf{v} = 0, \qquad \frac{\partial p}{\partial t} + b\bar{\rho}w = c^2 \frac{\partial \rho}{\partial t}, \tag{5.2}$$

where $c^2 = \kappa RT$ is the square of the sound speed, and $b = (\kappa - 1)g + \partial c^2/\partial z$ is the stratification parameter. The requirement that the vertical flow of mass $\bar{\rho}w$ go to zero at the boundaries of the atmosphere (at $z = 0$ and $z \to \infty$) serves as the natural boundary condition with respect to z for equations (5.1) and (5.2). The system (5.1), (5.2) is of fifth order with respect to time; for its unique solution it is necessary to specify initial values $\mathbf{v}_0$, p_0, and ρ_0 for all the unknown functions at $t = 0$.

For now,[12] let us suppose that $l = $ const (this is permissible for regions that are not too widely extended in latitude). Equations (5.1) and (5.2) will then have a family of stationary solutions $\mathbf{v}_s$, p_s, ρ_s, which describe *motions of the first kind*. Such motions are

(G_1) quasi-static (i.e., $\partial p_s/\partial z = -g\rho_s$),

(G_2) horizontal (i.e., $w_s = 0$),

and

(G_3) geostrophic (i.e., $u_s = -(1/l\bar{\rho})\,\partial p_s/\partial y, v_s = (1/l\bar{\rho})\,\partial p_s/\partial x$;

the latter equation indicates that the velocity field is nondivergent and that its stream function ψ_s equals $p_s/l\bar{\rho}$). Note that our equations have two invariants (stationary combinations of unknown functions):

$$ J_1 = (p - c^2\rho)_{z=0}, \quad J_2 = \bar{\rho}\left(\frac{\partial v}{\partial x} - \frac{\partial u}{\partial y}\right) + l\left[\frac{\partial}{\partial z}\left(\frac{p - c^2\rho}{b}\right) - \rho\right]. \quad (5.3) $$

The first of these is the linearized form of the entropy (at $z = 0$), and the second is that of the potential vorticity. It is possible to point out one integral invariant from which the invariance of both quantities J_1 and J_2 follows; it was discovered for the quasi-static case in 1958.[13]

Note also that there is an energy invariant

$$ J_3 = \int\left[\frac{\bar{\rho}v^2}{2} + \frac{p^2}{2\kappa\bar{p}} + \frac{g(p - c^2\rho)^2}{2\kappa\bar{p}b}\right]dV, \quad (5.4) $$

where the first term in the square brackets corresponds to the kinetic energy, the second to the elastic energy, and the third to the so-called thermobaric[14] energy (which is related to the forces of buoyancy acting on a particle displaced vertically from a state of equilibrium).

An arbitrary solution of equations (5.1) and (5.2) will be stationary if and only if the initial values have the properties (G_1) through (G_3). Otherwise, it will be the sum of a stationary solution with invariants (5.3) that are determined from the initial data (the invariants J_1 and J_2 and the conditions (G_1) through (G_3)

completely determine the stationary solution) and some non-stationary solution for which $J_1 \equiv J_2 \equiv 0$. Such nonstationary solutions describe *motions of the second kind*. These solutions are super-positions of waves of the form $\Phi(z) \exp[i(k_1 x + k_2 y - \sigma t)]$; because of their adiabatic character and their linearity, their frequencies σ are real. From now on, for simplicity let us consider the particular case $c^2 = \text{const}$, $b = \text{const}$ (this condition is exact for an isothermal atmosphere). In this case there are two types of waves: *two-dimensional waves*, in which there are no vertical oscillations of the air particles, that is $w = 0$; and *internal waves*, in which $w \neq 0$.

It can be verified that the frequencies of the two-dimensional waves are given by the formula $\sigma^2 = l^2 + k^2 c^2$ (where $k = (k_1^2 + k_2^2)^{1/2}$ is the horizontal wave number), so that the wave-front velocity equals the sound speed c; their amplitudes decrease mono-tonically with the altitude according to the equation $\Phi \sim (\bar{p})^{1/\kappa}$. Only waves of this sort are possible in a quasi-static barotropic atmosphere (Obukhov[15]); in the case of a quasi-static baroclinic atmosphere, they are dealt with in reference 13. The amplitudes of the internal waves depend on the altitude according to the formula $\exp[-z(b + g)/(2c^2) + imz]$, where $m \neq 0$ is the vertical wave number and their frequencies are given by the equation

$$(\sigma^2 - l^2)\left[\sigma^2 - \frac{(b + g)^2}{4c^2} - m^2 c^2\right] = k^2 c^2 \left(\sigma^2 - \frac{bg}{c^2}\right); \qquad (5.5)$$

for arbitrary k and m they fill the intervals $l^2 \leq \sigma^2 < bg/c^2$ and $\sigma^2 < (b + g)^2/4c^2$, corresponding to two different types of internal waves. Considering the case $l = 0$ in an isothermal atmosphere $(c^2 = \kappa g H, b = [\kappa - 1]g$, where H is the thickness of a homogeneous atmosphere), we can see that in the limit as $\kappa \to \infty$, as we reach isopycnic processes (i.e., an incompressible fluid), the second interval goes to infinity, and the first takes the form $\sigma^2 < g/H$, corresponding to internal gravitational waves. In the limiting case (as $\kappa \to 1$) of isothermal processes (in which the isothermal stratification becomes indifferent), the first interval vanishes into the point $\sigma^2 = 0$, and the second takes the form $\sigma^2 > g/4H$, corresponding to acoustic waves. Since $bg < (b + g)^2/4$, there is no overlap in the frequency spectra of acoustic waves and gravitational waves in an isothermal atmosphere. Diky[16] studied the wave spectra in a temperature-stratified (and therefore nonisothermal) atmosphere and established that only a very small overlap in acoustic and gravity-wave spectra occur in it. These results explain the presence of the minimum in the spectrum of Fig. 2.

In the quasi-static approximation—i.e., ignoring the left side of the third equation in (5.1) and consequently using the static equation $\partial p/\partial z = -\rho g$—all frequencies of the internal acoustic waves go to infinity; that is, they are completely "filtered out." The frequencies of the gravity waves are somewhat overestimated in this case; but the error is less for smaller k, i.e., in longer waves. The frequencies of the two-dimensional waves, the stationary solutions, and the invariants do not change in this case; using the quasi-static approximation to describe the synoptic processes is thereby justified.

In allowing for the curvature of the earth, the foregoing results undergo some alteration. The most important change is a transformation of the stationary solutions (motions of the first kind) into slow *gyroscopic waves*. It is possible to allow approximately for the curvature of the earth in the following manner: Using the first two equations of (5.1), construct equations for the vorticity $\partial v/\partial x - \partial u/\partial y$ and the divergence $\partial u/\partial x + \partial v/\partial y$, allowing for the dependence of the Coriolis parameter l on the coordinate along the meridian y. In the equations obtained, replace l and $\beta \equiv \partial l/\partial y = 2\omega \sin\theta/a$ by constants (that is, transform into the so-called β-plane). Then, in the case of a barotropic atmosphere and further, ignoring its horizontal compressibility (i.e., assuming that $\partial u/\partial x + \partial v/\partial y = 0$), we obtain the equation $\sigma = -\beta k_1/k^2$ for the frequencies of the gyroscopic waves (Rossby[17]). The minus sign indicates that the waves move toward the west.

Making an exact allowance for the sphericity of the earth, Hough,[18] as early as the end of the nineteenth century, established the distinction between the two kinds of waves in a barotropic atmosphere: the slow gyroscopic waves, which are important for forecasting the weather, and the fast two-dimensional waves, which are important for describing the tides (see also the works on tide theory by Love,[19] Kochin,[20] Pekeris,[21] Kertz,[22] and Siebert[23]). Here the effect of the weak horizontal compressibility can be described by using a power series in the parameter $\gamma = (2\omega a/c)^2$ (Yaglom[24]). As $\gamma \to 0$, the fast waves disappear, and the frequencies of the gyroscopic waves are given by the equation

$$\sigma = \alpha m - \frac{2(\alpha + \omega)m}{n(n+1)}, \tag{5.6}$$

where m and n are integers and α is the above-mentioned circulation index (Haurwitz,[25] Blinova[26]). For the earth's atmosphere ($\gamma \approx 10$) this equation is rather crude when n and m are small, but its accuracy increases rapidly with increasing n and m.

Diky[27,28] described all the possible types of waves for a baroclinic atmosphere on a spherical earth. His earlier paper [27] dealt with the isothermal atmosphere, and the later one[28] dealt with the so-called standard atmosphere CIRA-1961. Seeking waves of the form $\Phi(\zeta)\Psi(\cos\theta)\exp[i(m\lambda + \sigma t)]$ (where λ is the longitude, θ is the colatitude, and $\zeta = \ln(p/p_0)$ is the vertical coordinate), Diky obtained for Ψ the so-called Laplace equation of tidal theory. This equation contains m, σ, and the constant h (the so-called depth of the dynamically equivalent ocean), which arises in the separation of the variables; but it does not contain the stratification characteristics. For Φ he obtained a "vertical" equation containing σ, h, and the stratification characteristics, but not the horizontal wave number m. Each of these equations makes it possible to determine a family of characteristic curves $\sigma(h)$; the intersections of the curves of different families determine the possible eigenvalues of σ and h.

Figure 8 shows the "horizontal" characteristic curves $\sigma(h)$ for $m = 2$; the curves in the lower portion correspond to the fast waves, and those in the upper half, to the slow waves. Figure 9 gives the "vertical" characteristic curves $\sigma(h)$; those on the lower right correspond to acoustic waves, and those on the upper left correspond to gravitational waves; the numbers with the curves indicate the number of nodes of the corresponding eigenfunctions.

6. Adaptation of Meteorological Fields

Let us return for a moment to the simplified model of the atmosphere considered above: the "flat" model with $l = $ const. As was already noted, if the initial values $\mathbf{v}_0$, p_0, ρ_0 have properties (G_1) through (G_3), i.e., if only motions of the first kind are present at the initial instant, then only these motions will remain in the future (since the solutions describing them are stationary). If, on the other hand, conditions (G_1) through (G_3) are violated at the initial instant in some region of space V, then in that region there will also be motions of the second kind—fast waves. However, these waves scatter in every direction; and when they have left region V, conditions (G_1) through (G_3) prevail—that is, only some motions of the first kind remain (defined by the invariant fields J_1 and J_2, which can be constructed from the initial data). This process of restoring the consistency conditions (G_1) through (G_3) of the meteorological fields $\mathbf{v}$, p, ρ is called the *adaptation* of meteorological fields. The problem of the adaptation of meteorological fields in the case of a quasi-static barotropic atmosphere was first formulated by Rossby[29] and Cahn[30] and solved by Obukhov;[15] for the baroclinic atmosphere, this problem was treated by Bolin,[31] Kibel'[32] (without taking the two-dimensional waves into account), Veronis,[33] Fjelstad,[34] and Monin;[13] see also the excellent review by Phillips[35] dealing with geostrophic motions (or, in our terminology, motions of the first kind) in the atmosphere and ocean.

   Adaptation to the state of static equilibrium (G_1) is brought about by the generation and scattering of internal acoustic waves. The

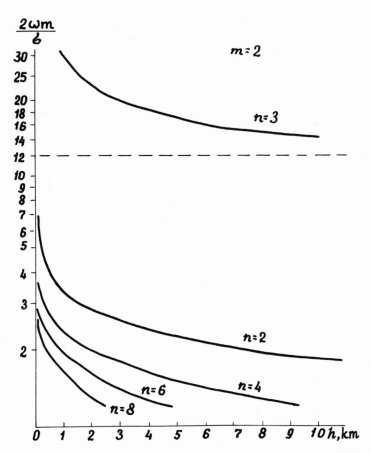

Fig. 8   The characteristic curves $\sigma(h)$ of the Laplace equation of tide theory for $m = 2$. After Diky.[27]

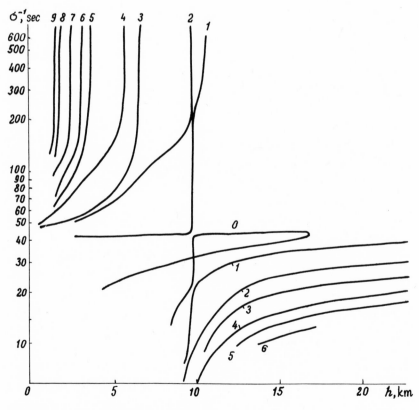

Fig. 9   The characteristic curves $\sigma(h)$ of the "vertical" equation for a standard atmosphere CIRA–1961. After Diky.[27]

duration of this process is approximately the same as the time required for a front of internal acoustic waves to traverse (with the sound speed $c \sim 20$ km/min) the main thickness of the atmosphere—a process requiring only a few minutes in all. After this, the atmosphere continues to adapt to the state of geostrophic equilibrium (G_2) and (G_3). On the average over the thickness of the atmosphere, this state is reached after the two-dimensional waves (whose fronts move with the same speed c) have left region V. It is reached still later for all altitudes—after the slower internal gravitational waves have dispersed (their speeds depend on the thermal stratification of the atmosphere; [13,32,36] the fronts move with the speed $2[(1 - 1/\kappa)(1 - r/r_0)RT]^{1/2}$, and behind the fronts, as in the case of two-dimensional waves, there is a continuous "wake" in which damped oscillations take place).

Figure 10 gives an example of the adaptation of meteorological fields. In this case, there were no pressure perturbations at the initial instant of time, and the velocity field corresponded to a plane-parallel flow of the type of tangential discontinuity along the ordinate axis (the initial distribution of the surface velocity $\mathbf{v}_0(x)$ is given by the dotted line in Fig. 10). The velocity field changed only slightly as a result of adaptation; see the limiting distribution of the surface velocity $\mathbf{v}(x)$ (the kinetic energy decreased by 3% from losses due to the generation of fast waves and the formation of inhomogeneities in the pressure field). The pressure field actively "adapted" to the velocity field: a distinct dip was produced in it (see the limiting distribution of the altitudes of the isobaric surface $z(x)$ at ground level; it dropped by 4 dkm along the ordinate axis).

So far we have dealt only with the adaptation of meteorological fields in a "flat" model of the atmosphere with $l =$ const. Allowing for the sphericity of the earth's surface introduces two relatively insignificant changes into this process. First, the fast waves escaping from the perturbed region will now propagate, not in an infinite space, but in one that is horizontally bounded; this will produce an interference pattern that is attenuated with time because of dissipative processes (which were not taken into account above). Second, motions of the first kind will not be stationary on the sphere but will be a superposition of slow gyroscopic waves. The motion of these waves will at all times lead to changes in the configuration of the fields θ and Ω. These changes in turn upset the consistency conditions (G_1) through (G_3) of the meteorological fields. Thus two competing processes continually take place on a sphere: (1) disturbance of the consistency of the velocity, pressure, and density fields as a result of the evolution of the spatial distribution of the entropy and the potential vorticity and (2) adaptation of the meteorological fields as a result of the generation, dispersion, and damping of the fast waves.

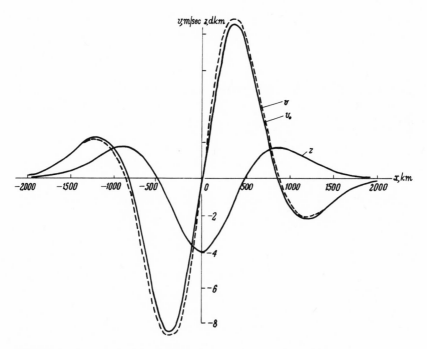

Fig. 10 An example of the adaptation of meteorological fields in the baroclinic atmosphere. After Monin.[13]

So far in this section, as in the preceding one, we have considered
only the waves of small amplitude; these can be described by the
linearized dynamic equations. In a real atmosphere, motions of the
second kind—acoustic and gravitational waves—in fact nearly always have very small amplitudes (and for this reason, they produce the "meteorological noise" that has little importance for the
weather); motions of the first kind, however, no longer have small
amplitudes, generally speaking, and they are described by the nonlinear equations (4.1). This nonlinearity (as well as allowance for
the sphericity of the earth's surface) results in the nonstationary
character of motions of the first kind—an evolution of the spatial
distribution of the quantities θ and Ω as a result of their being
transported by air currents. As a consequence, there is a continuous
competition between the disturbance of consistency and the adaptation of the meteorological fields. Because of this competition,
disturbances of the consistency conditions (G_1) through (G_3) turn
out to be small, as a rule, and the motions of the first kind still
satisfy these conditions, albeit approximately. This statement will
be made more precise in the next section.

7. The Quasi-geostrophic Approximation

It is desirable to simplify the hydrodynamic equations in such a way
that the simplified equations describe the motions of the first kind,
which are important for the weather, with sufficient accuracy; but
they should not contain the unimportant motions of the second kind
in their solutions (i.e., the latter are "filtered out"). As was already
stated, condition (G_1), which requires that motions of the first kind
be quasi-static, is exactly satisfied. By using it instead of the complete
equation for motion along the vertical, acoustic waves are "filtered
out" of the solutions to the hydrodynamic equations. Later we shall
be using this "quasi-static approximation" everywhere. With its
aid we can use the pressure p as the vertical coordinate in place of
the altitude z. Once this is done, the pressure field $p(x, y, z, t)$ will
be replaced in the hydrodynamic equations by another unknown

function $z(x,y,p,t)$, the values of which are, for fixed p, the altitudes of the isobaric surfaces $p = $ const.

In changing from the coordinates (x,y,z) to (x,y,p), the horizontal pressure gradient $\nabla_h p$ is replaced by $\rho g \nabla_h z$ (and $\partial p/\partial t$ is replaced by $\rho g\, \partial z/\partial t$); the static equation $\partial p/\partial z = -\rho g$ is rewritten in the form $\partial z/\partial p = -1/g\rho$. When ρ has been determined from this equation, Clapeyron's equation $p = \rho R T$ can be put into the form $T = -(g/R)p(\partial z/\partial p)$. The individual derivative d/dt takes the form $d_h/dt + w^*(\partial/\partial p)$, where d_h/dt is the derivative with respect to the horizontal motion (on isobaric surfaces) and $w^* = dp/dt$ replaces the magnitude of the vertical velocity. One of the advantages of using the coordinates (x,y,p) is the particularly simple form of the continuity equation $D + \partial w^*/\partial p = 0$, where $D = \partial u/\partial x + \partial v/\partial y$ is the horizontal divergence of the velocity. On the other hand, the boundary condition $w = 0$ at $z = 0$ assumes the more complicated form $w^* = \rho g\, d_h z/dt$ (for simplicity, it is usually not required that this condition be satisfied at $z = 0$, but at $p = p_0$, where p_0 is the average surface pressure).

The hydrodynamic equations (or more accurately, the Eulerian equations) contain two dimensional parameters, the acceleration due to gravity g and the Coriolis parameter l (on a sphere it is better to use the curl of the earth's rotation 2ω; the earth's radius a also figures in the equations on a sphere). The average surface pressure p_0 and the density ρ_0 are also included in the boundary condition at the surface of the earth [by using these data it is possible to define the height of the homogeneous atmosphere $H = p_0/\rho_0 g$ and the isothermal sound speed $c_0 = (gH)^{1/2}$]. Let L and U be the typical length and speed scales for the synoptic processes. From the foregoing dimensional quantities it is possible to set up the following four dimensionless parameters:[37-39] (1) the *sphericity parameter* L/a; (2) the *quasi-static parameter* H/L; (3) the *Kibel number* $\mathrm{Ki} = U/Ll$ (in some foreign papers it is also sometimes called the Rossby number); (4) the *Mach number* $\mathrm{Ma} = U/c_0$ (or the parameter of the horizontal compressibility of the atmosphere $L/L_0 = \mathrm{Ma}/\mathrm{Ki}$, where $L_0 = c_0/l$ is the above-mentioned scale of the oscillations of a two-dimensional compressible atmosphere in the field of the Coriolis force, introduced by Obukhov[15]).

The Kibel number Ki can be interpreted as the ratio of the typical value of U/L for the relative vorticity Ω_z to the angular velocity of the earth's rotation $2\omega_z = l$ (or as the ratio of the typical relative acceleration U^2/L to the typical Coriolis acceleration Ul). With the exception of the tropical zone, this number as a rule is small; thus, according to calculations by Chaplygina,[40] who determined the values of Ω_z/l from real data, the magnitude of this number is almost always smaller than 0.4, and in 75% of the cases it is smaller than 0.2. This means that the rotation of the air in large-scale atmospheric vortices at moderate and high latitudes (cyclones and anticyclones) takes place significantly more slowly than the rotation of the earth. Consequently, in the scales of the synoptic processes, the horizontal pressure gradient is approximately balanced by the Coriolis force. In other words, the geostrophy conditions (G_3) are approximately satisfied (with accuracy to terms on the order of $U \cdot \mathrm{Ki}$), and in the coordinates (x, y, p) they take the form

$$u = -\frac{g}{l}\frac{\partial z}{\partial y}, \qquad v = \frac{g}{l}\frac{\partial z}{\partial x}. \tag{7.1}$$

From (7.1) it follows that in the scales of the synoptic processes, variations of the quantity z in the horizontal direction are on the order of lLU/g (the time variations of z are of the same order). By using (4.3) and the static equation $1/\rho = -g\,\partial z/\partial p$, the adiabaticity equation $ds/dt = 0$ can be put into the form

$$gp^2\frac{d_h}{dt}\frac{\partial z}{\partial p} + \alpha_0^2 c_0^2 w^* = 0, \tag{7.2}$$

where $\alpha_0^2 = -(T/T_0)(p/c_p)(\partial s/\partial p)$ is the *dimensionless parameter of static stability* (the quantity $\mathrm{Ri} = (1/\mathrm{Ma}^2)(T_0^2/T^2)\alpha_0^2$ is sometimes called [37,38] the *Richardson number*). From equation (7.2) it can be seen that the values of w^* are on the order of $(p_0 U/L)(L^2/L_0^2\alpha_0^2)\,\mathrm{Ki} \approx (p_0 U/L)\,\mathrm{Ki}$ (in the troposphere, usually $\alpha_0^2 \sim L^2/L_0^2$). Finally, the continuity equation $D + \partial w^*/\partial p = 0$ shows that the values of the horizontal divergence of the velocity are of the order $w^*/p_0 \sim$

(U/L) Ki. Thus, for motions of the first kind (the synoptic processes) the following conditions are satisfied:

$$\left.\begin{aligned}
\Omega_z &\equiv \frac{\partial v}{\partial x} - \frac{\partial u}{\partial y} = \frac{g}{l}\nabla^2 z + O\left(\frac{U}{L}\text{Ki}\right), \\
D &\equiv \frac{\partial u}{\partial x} + \frac{\partial v}{\partial y} = O\left(\frac{U}{L}\text{Ki}\right),
\end{aligned}\right\} \tag{7.3}$$

where $\nabla^2 = \partial^2/\partial x^2 + \partial^2/\partial y^2$. On the contrary, for motions of the second kind these conditions are not satisfied. Therefore it is possible to "filter out" the fast waves by finding solutions of the hydrodynamic equations in the form of asymptotic series in powers of Ki with principal terms satisfying conditions (7.3) (the so-called quasi-geostrophic expansion, first proposed by Kibel'[41]). The equations of these principal terms describe the synoptic processes with sufficient accuracy (with a relative error only on the order of Ki) and do not contain fast waves among their solutions. Such equations are relations (7.1) and the equation obtained from the conservation law $d_h\Omega/dt + w^* \, \partial\Omega/\partial p = 0$ of the potential vorticity $\Omega \approx -\Omega_{az}g\partial s/\partial p$ after eliminating from it, with the aid of (7.2), the quantity w^* and considering only terms of zeroth order in Ki. The latter equation is reduced to the elegant form

$$\mathfrak{F}\frac{\partial z}{\partial t} = -\frac{g}{l}[z, \mathfrak{F}z + l], \tag{7.4}$$

where $\mathfrak{F}$ is an elliptic linear operator (the analog of the three dimensional Laplace operator) defined by the formula

$$\mathfrak{F}z = \frac{g}{l}\nabla^2 z + \frac{gl}{c_0^2}\frac{\partial}{\partial p}\frac{p^2}{\alpha_0^2}\frac{\partial z}{\partial p}, \tag{7.5}$$

and the square brackets $[A,B]$ will henceforth denote the Jacobian $\partial(A,B)/\partial(x,y)$. Equation (7.4) is obtained from the more general equation

$$\frac{d_h}{dt}\left(\ln\Omega_{az} + \frac{g}{c_0{}^2}\frac{\partial}{\partial p}\frac{p^2}{\alpha_0{}^2}\frac{\partial z}{\partial p}\right) = 0, \tag{7.4'}$$

which expresses a certain approximate conservation law for horizontal motions.

The method of asymptotic expansions, which makes it possible to derive equation (7.4) from the initial hydrodynamic equations, is a particular case of the general asymptotic methods developed by Bogolyubov and Krylov. These were developed for the description of slow oscillations in nonlinear mechanical systems in which rapid oscillations occur along with the slow ones (Van der Pol was the first to develop one such method for the description of current oscillations in an electric circuit containing a vacuum tube with feedback). From a purely mathematical point of view, we are dealing here with equations (describing the oscillations) that contain a small parameter (the Kibel number) in the terms containing higher derivatives (terms in the equations of motion describing the relative accelerations).

Equation (7.4) contains only one unknown function, z. It describes synoptic changes in the three-dimensional field of atmospheric pressure valid for the quasi-geostrophic approximation. The equation is first-order with respect to time; this is natural since the original system of hydrodynamic equations was fifth-order, and the two families of waves (acoustic and gravitational) that were filtered out each reduced the order by two. Thus, to predict the field of the atmosphere pressure in the quasi-geostrophic approximation, it is sufficient to know the initial values of only the pressure field itself; the initial values of the velocity field (which would be needed to solve the complete hydrodynamic equations) need no longer be known. This simplification is very important in practice since the wind field is presently measured rather crudely, and providing for its exact measurement would be a very cumbersome and expensive matter.

The differential equation (7.4) is of second order in p, and for its solution it is necessary to specify the boundary conditions at the upper and lower limits of the atmosphere $p = 0$ and $p = p_0$. For $p \to 0$ we require that the kinetic energy be bounded, i.e., $p|\nabla_h z|^2 < \infty$; at $p = p_0$ we use the condition $w = 0$, which, with the aid of equation (7.2) at $w^* = \rho_0 g \, d_h z / dt$ (see the small print on p. 33), reduces to the form $d_h / dt \, (p \partial z / \partial p + \alpha_0^2 z) = 0$. With these boundary conditions equation (7.4) can be rewritten in the integral form

$$\left(\mathfrak{G} \nabla^2 - \frac{1}{L_0^2} \right) \frac{\partial z}{\partial t} = - \mathfrak{G} \left[z, \frac{g}{l} \nabla^2 z + l \right] - \frac{1}{L_0^2} \frac{g}{l} \int\limits_{p}^{p_0} \left[z, \frac{\partial z}{\partial p} \right] dp, \quad (7.6)$$

where $\mathfrak{G}$ is the operator of integration with respect to p defined by the equation

$$\mathfrak{G} z = \frac{1}{p_0} \int\limits_{0}^{p_0} z \, dp + \int\limits_{p}^{p_0} \frac{\alpha_0^2 \, dp}{p^2} \int\limits_{0}^{p} z \, dp. \quad (7.7)$$

Equation (7.6) is particularly convenient for the transition in the limit as $\alpha_0^2 \to 0$ to the case of the barotropic atmosphere. For the synoptic processes in a barotropic atmosphere, the magnitude of z is the sum of some standard function of p and the product of a function of x, y, and a function of p (see my paper[42]); the second term in the right side of (7.6) drops out, and equation (7.6) takes the simple form

$$\left(\nabla^2 - \frac{1}{L_0^2} \right) \frac{\partial z}{\partial t} = - \left[z, \frac{g}{l} \nabla^2 z + l \right]. \quad (7.6')$$

This equation was proposed for the purposes of weather forecasting by Obukhov[15] and Charney[4, 43] (using a different method, Ertel[44] earlier obtained a similar equation, only without the term with $1/L_0^2$; but the significance of that equation was not understood at the time). This is precisely the equation that was first used in the postwar years (for example in a paper by Charney, Fjørtoft, and von Neumann[45]) for numerical forecasts of the pressure field (at some middle level in the troposphere).

From (7.6′) it follows that the value of $\partial z/\partial t$ at a fixed point M is obtained by integrating the "advection of the vorticity" $[z, (g/l)\, \nabla^2 z + l]$ over all points of the plane M' with a weighting factor $(1/2\pi)\, K_0\,(r/L_0)$, where r is the distance from M' to M, and K_0 is the symbol for the cylindrical MacDonald function. The influence function $K_0\,(r/L_0)$ decreases with increasing r; thus, changes of the pressure $\partial z/\partial t$ at each fixed point M are determined by the entire pressure field $z(M')$, but the influence of distant points M' turns out to be small (the distance L_0 serves as the "influence radius"). Note that it is essential to retain $1/L_0^2$ in the operator $\nabla^2 - 1/L_0^2$ in the left side of equation (7.6′) [and similarly, in the operator $\mathfrak{G}\nabla^2 - 1/L_0^2$ in the left side of equation (7.6)]: If this subtrahend (which describes the effect of the "horizontal compressibility" of the atmosphere) were ignored, the influence function $K_0(r/L_0)$ would be replaced by the function $\ln r$, which increases with increasing r, i.e., corresponds to an increasing influence of distant points M' with increasing distance r, which of course is not natural. If the horizontal field $z(x,y)$ is represented in the form of a superposition of elementary harmonic waves, then it becomes clear that for waves with lengths much shorter than L_0, allowance for the term $-1/L_0^2$ in the dynamic operator is insignificant, but it becomes quite appreciable for the description of the evolution of long waves (with length $L \gtrsim L_0$).

In analogy with the foregoing, it follows from (7.6) that the value of $\partial z/\partial t$ at a fixed point $M(x,y,p)$ of a baroclinic atmosphere is obtained by integrating—over all points $M'(x',y',p')$—the sum of the "vorticity advection" $[z, (g/l)\nabla^2 z + l]$ with some weight $G(r/L_0; p,p')$ and the "heat advection" $-g/l\,[z,p\,\partial z/\partial p]$ with weight $(1/L_0^2\alpha_0^2)$

$p'(\partial/\partial p')\, G(r/L_0; p, p')$, where r is the horizontal distance from M' to M. The influence functions G and $p'\, \partial G/\partial p'$ were first determined (using coordinates x, y, z and Fourier transforms in x and y) in 1951–1952 by Obukhov and Chaplygina[46] and almost simultaneously (using coordinates x, y, p) by Buleyev and Marchuk,[47] and later by Hinkelmann[48] and Kuo.[49] The simplest derivation of these functions is given in one of my papers,[13] where the role of these influence functions in the problem of the adaptation of meteorological fields is also established. Figures 11 and 12 show plots of these functions from the paper by Obukhov and Chaplygina[46] (in dimensionless form, after Fourier transformations with respect to x and y at a value of the dimensionless wave number $kL_0\{[(\gamma_a - \gamma)/(\kappa - 1)]\,(\kappa R/g)\}^{1/2} = 4$), showing clearly the relative weights with which the "dynamic" and "thermal" contributions of different layers of air enter into the values of $\partial z/\partial t$ at various levels.

8. The Quasi-solenoidal Approximation

As the equator is approached, the Coriolis parameter $l = 2\omega \cos\theta$ decreases, the Kibel number $\mathrm{Ki} = U/Ll$ ceases to be small, and consequently the quasi-geostrophic approximation loses its validity. Moreover, experience with numerical forecasts of the pressure field has shown that even outside the tropical zone, describing the synoptic processes by means of the quasi-geostrophic approximation is not sufficiently accurate in some cases. Therefore, in place of conditions (7.3), it may be useful to find other "consistency" conditions for the synoptic fields of the velocity and pressure that will make it possible to distinguish the synoptic processes from the fast wave motions.

Such conditions (suitable not only near the equator but everywhere) can be obtained by using the following fact as a starting point: In the horizontal velocity field of the slow synoptic motions, the potential component is small compared to the solenoidal component. In other words, the horizontal divergence of the velocity $D = \partial u/\partial x + \partial v/\partial y$ is small (in absolute magnitude) in comparison

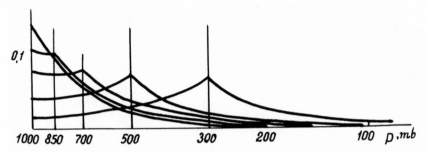

Fig. 11   The influence functions of vorticity advection at various air levels, showing the relative importance of the "dynamic" contribution to the values of $\partial z/\partial t$ at levels $p = 1000$, 850, 700, 500, and 300 mb. After Obukhov and Chaplygina.[46]

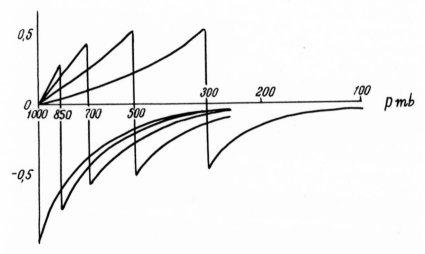

Fig. 12   The influence functions of heat advection at various air levels, showing the relative importance of the "thermal" contribution to the values of $\partial z/\partial t$ at levels $p = 1000$, 850, 700, 500, and 300 mb. After Obukhov and Chaplygina.[46]

to the vorticity $\Omega_z = \partial v/\partial x - \partial u/\partial y$. As a result of this, in the equation for D obtained by applying the divergence operation to the equations of motion, the principal terms will be those that contain neither D nor $w^* \sim p_0 D$. If the term containing the Coriolis parameter as a factor is ignored, then the principal terms are of the form $2[u,v] - g\nabla^2 z$; from a comparison of these terms it follows that the variations of z are on the order of $U^2/g = H \cdot \mathrm{Ma}^2$, where $\mathrm{Ma} = U/c_0$ is the Mach number, which is quite small for the synoptic processes. Then from (7.2) we obtain $w^* \sim p_0 (U/L) (\mathrm{Ma}^2/\alpha_0^2)$ and therefore $D \sim (U/L) (\mathrm{Ma}^2/\alpha_0^2)$ as well. In other words, the condition that D be small compared to Ω_z in motions of the first kind (the synoptic processes) can be written in the form

$$\Omega_z = O\left(\frac{U}{L}\right), \qquad D = O\left(\frac{U}{L}\frac{\mathrm{Ma}^2}{\alpha_0^2}\right). \tag{8.1}$$

For motions of the second kind (fast waves), these conditions are, on the contrary, not satisfied. Therefore, it is possible to "filter out" the fast waves by finding solutions of the hydrodynamic equations in the form of asymptotic power series in $\mathrm{Ma}^2 = U^2/c_0^2$ with principal terms satisfying conditions (8.1) and with a principal term for z on the order of $H \cdot \mathrm{Ma}^2$. The principal component of the wind-velocity field (u,v) will then be its *solenoidal* component, and $u = -\partial\psi/\partial y$ and $v = \partial\psi/\partial x$, where ψ is the stream function; for this reason the asymptotic series indicated are sometimes called the *quasi-solenoidal expansion*. One of the equations for the principal terms of the quasi-solenoidal expansion is obtained by considering only the terms of zeroth order in Ma in the approximate conservation law (7.4′):

$$\frac{\partial F}{\partial t} = -[\psi,F], \qquad F = \ln (\nabla^2\psi + l) + \frac{g}{c_0^2}\frac{\partial}{\partial p}\frac{p^2}{\alpha_0^2}\frac{\partial z}{\partial p}. \tag{8.2}$$

Unlike the quasi-geostrophic approximation (7.4), which contains only one unknown function, z, equation (8.2) contains two unknown functions, z and ψ. The connection between them [i.e., the relationship between the velocity and pressure fields, which was given in the geostrophic approximation by equations (7.1)] will be given by an equation that is obtained from the above-mentioned equation for D in which only the terms of zeroth order in Ma are retained. This equation, called the balance equation, has the form

$$g\nabla^2 z = (\nabla \cdot l\nabla)\psi + 2\left[\frac{\partial\psi}{\partial x}, \frac{\partial\psi}{\partial y}\right].$$ (8.3)

Equations (8.2) and (8.3) describe the synoptic changes of the velocity and pressure fields in the quasi-solenoidal approximation. They are of first order with respect to time (since the two "filtered out" families of fast waves, acoustic and gravitational, each reduced the order by two). Thus, for forecasting the synoptic processes under the quasi-solenoidal approximation, it is sufficient to have only the initial values of the pressure field z; and the initial values of the velocity field (the stream function ψ) can be determined from the field z by using equation (8.3).

The equations of the quasi-solenoidal approximation, (8.2) and (8.3), are suitable both near the equator, where l is small and the quasi-geostrophic approximation loses its validity, and outside the tropic zone, i.e., in the regions where the Kibel number Ki is small. In the latter regions the quasi-geostrophic approximation is valid, with accuracy to terms of the order Ki; and the quasi-solenoidal approximation has even greater accuracy, to terms on the order of Ki^2 [here the second term in the right side of (8.3) will be smaller than the other terms by the factor Ki, and if it is ignored along with the changes in l with latitude, then the geostrophic relation $l\psi \approx gz$ is obtained, whereby (8.2) is transformed into the quasi-geostrophic equation (7.4)].

The balance equation (8.3) was obtained as a second approximation in the quasi-geostrophic expansion in one of my papers[42] (see also the papers by Bolin [50] and Thompson[51] and my later paper[13]); it was pointed out in Charney's paper[52] with reference to an unpublished paper by Fjørtoft. A justification of the quasi-solenoidal approximation by means of an asymptotic expansion in powers of Ma was presented in a paper of mine[36] and in a similar form by Charney;[37] Gavrilin[39] derived the equations of the quasi-solenoidal approximation for nonadiabatic synoptic processes on a spherical earth.

Actually, the quasi-solenoidal approximation was used a long time ago in describing the synoptic processes: The theory of synoptic waves developed by Blinova[26] (in the terminology of section 5, gyroscopic waves) was based on a consideration of the quasi-solenoidal approximation for the vortex transport equation in an adiabatic barotropic atmosphere on a spherical earth [in the case of a barotropic atmosphere, it is possible to set $\partial z / \partial p = 0$ in (8.2), and then F can be simply replaced by $\nabla^2 \psi + l$]. In the indicated theory this equation was linearized relative to a state in which the atmosphere rotates around the earth as a rigid body with angular velocity α (the circulation index mentioned above), and it took the form

$$\left(\frac{\partial}{\partial t} + \alpha \frac{\partial}{\partial \lambda} \right) \nabla^2 \psi + 2 \frac{\alpha + \omega}{a^2} \frac{\partial \psi}{\partial \lambda} = 0, \tag{8.2'}$$

where λ is the longitude and a is the radius of the earth (an analog of this equation in Cartesian coordinates was first proposed for the description of the gyroscopic waves by Rossby,[17] who obtained from it an equation for the speed of motion of baric depressions, i.e., planetary pressure waves; see section 5). The elementary wave solutions of equation (8.2') were pointed out by Haurwitz.[25] Using this equation, Blinova constructed the general solution of the initial-value problem for the atmospheric-pressure field, relating the latter with the field ψ by an equation that is in fact the linearized balance equation.

Solving the equations (8.2) and (8.3) of the quasi-solenoidal approximation entails serious mathematical difficulties (see Charney[39, 52]). In the first place, from the balance equation (8.3) it is necessary to find an initial field ψ for a given initial field z. Second, for each step of integration of (8.2) with respect to time, it is necessary to find a field ψ for the field F [which is determined by the second formula of (8.2)] by using the same balance equation (8.3). Equation (8.3), considered as an equation in ψ, is among the so-called Monge-Ampère equations. In practice, because of the incompleteness or even absence of synoptic information over a considerable fraction of the earth's surface, equation (8.3) must be solved with respect to ψ only within some bounded territory, specifying in some manner the values of ψ on its boundary. Such a boundary-value problem for equation (8.3) will be correct only if it is elliptical. The condition for its ellipticity reduces to the form $g\nabla^2 z + l^2/2 > 0$. One can be sure that such a condition is satisfied only when the values of Ki are small, but if Ki is not small (e.g., in the tropics), then this condition may not be satisfied. On the other hand, in the case of small Ki, equation (8.3) can be rewritten (in the simplest case with $l = $ const) in the form[13]

$$l\nabla^2 \psi = g\nabla^2 z - 2\frac{g^2}{l^2}\left[\frac{\partial z}{\partial x}, \frac{\partial z}{\partial y}\right]. \tag{8.3'}$$

9. Primitive Equations

In recent years, several papers on the hydrodynamic theory of short-range weather forecasts (for example, Smagorinsky,[53] Hinkelmann,[54] Phillips,[55] Charney,[37] Buleyev and Marchuk[56]) have displayed a tendency to forego "filtering out" the fast waves and to resume using the complete hydrodynamic equations (though in the quasi-static approximation), which have then been called "primitive," i.e., initial. (The first to suggest this was Kibel'; see the conclusion of his paper.[41]) This tendency can probably be explained by the following reasons: (1) the quasi-geostrophic approximation is not sufficiently accurate in many cases; (2) solution of the equations of the quasi-solenoidal approximation involves mathematical difficulties; (3) owing to the development of computational mathematics and computer techniques, numerical solution of the primitive equations is now perhaps no more difficult or only slightly more difficult than the solution of the filtered equations (see, for example, the finite difference schemes proposed by Marchuk[57-60] for the numerical solution of the primitive equations).

The primitive equations, just as the balance equation considered above, must be integrated within the limits of bounded territories. This raises the question of the correct formulation of the corresponding boundary conditions: For insufficient as well as superfluous boundary conditions the solutions of the equations will be unstable, and the errors produced in each step of the integration with respect to time will extend farther and farther from the boundaries and toward the interior of the territory under consideration. Charney[37] has shown that the correct boundary-value problem for the primitive equations is obtained by specifying the normal component of the velocity on the entire boundary of the territory considered and by specifying the values of the potential vorticity on those sections of the boundary where the air motion is directed toward the interior (similar equations were earlier formulated by Charney, Fjørtoft, and von Neumann[45] for the equation of the quasi-geostrophic approximation in a barotropic atmosphere).

The attempt to increase the accuracy of describing the synoptic processes (in comparison to the quasi-geostrophic approximation) by returning to the primitive, unfiltered equations is made at the cost, first, of retaining the high order with respect to time in the system of equations employed and consequently being forced to assign a large number of initial data (namely, the initial values of not only the pressure field but also the wind-velocity field) and, second, retaining gravitational waves among the solutions of the prognostic equations.

As was noted in section 6, the fronts of two-dimensional gravitational waves travel at the sound speed c_0, and internal gravitational waves travel with the speed $2\alpha_0 c_0$, where α_0^2 is the parameter of static equilibrium, which was introduced in (7.2); at the same time, the phase velocities of internal gravitational waves can have values ranging from 0 to $2\alpha_0 c_0$, i.e., among these waves they can be arbitrarily slow.[36] It would not be expedient to filter out such slow waves if they were able to have significantly large amplitudes, i.e., if they could be important for the weather. However, observational data indicate that gravitational waves of large amplitude evidently occur only very rarely in the atmosphere. Therefore, in using the primitive equations, gravitational waves should be considered only as hindrances that one desires, in at least some way, to filter out.

Recall that gravitational waves can be generated, first, as a result of an initial inconsistency between the pressure and velocity fields and, second, because the nonstationary character of the synoptic processes (due to their nonlinearity and also to the influence of the sphericity of the earth) constantly disturbs the consistency between these fields. When the primitive equations are used, the second of these factors remains in force; but the first can be eliminated by specifying (according to actual data) only the initial pressure field and choosing the initial velocity field consistently with the pressure field (this eliminates at once all the errors connected with the inaccuracy of measuring the initial wind field; see Hinkelmann[61]). These conditions can be written, in accordance with the foregoing, in terms of the quasi-geostrophic or the quasi-solenoidal expansion (see my paper[36] and the article by Phillips[55]). With accuracy to terms of order Ki^2, these conditions reduce to the requirement that the stream function ψ must be related to z by the balance equation (8.3) and that the velocity divergence $D = \partial u/\partial x + \partial v/\partial y$ be related to z by the formula

$$D = -\frac{g}{l^2}\left\{\nabla^2\frac{\partial z}{\partial t} + \left[z, \frac{g}{l}\nabla^2 z + l\right]\right\}, \qquad (9.1)$$

in which it is necessary in addition to express $\partial z/\partial t$ in terms of the values of the field z at the same instant in time with the aid of equation (7.4) or equation (7.6).

And so, the advantage of returning to the primitive equations has no connection to taking the gravitational waves into account; these, on the contrary, must be filtered out (at least in part, by making the initial velocity and pressure fields approximately "consistent"). But its usefulness may lie in its allowing for the actual boundedness of the "interaction radius" of the baric field, which follows from the fact that the system of the primitive equations is hyperbolic: The baric tendency, i.e., the derivative $\partial z/\partial t$, enters in the main linear part of these equations under the sign of the hyperbolic operator $\mathfrak{G}\nabla^2 - 1/L_0^2 - (1/c_0^2)\,(\partial^2/\partial t^2)$, where $\mathfrak{G}$ is defined by formula (7.7). Therefore the value of $\partial z/\partial t$ at a fixed point M at the instant of time t is determined by the values of the initial pressure and velocity fields in a horizontally bounded vicinity of point M, with a radius $c_0 t$; it cannot in fact depend on the values of the initial fields outside this vicinity. In the filtered equation, on the other hand, (7.6), for example, this hyperbolic operator is replaced by the elliptic operator $\mathfrak{G}\nabla^2 - 1/L_0^2$; as a result, the value of $\partial z/\partial t$ at point M at any instant of time t becomes dependent on the values of the initial pressure field over all space, including the values of this field at points located at a distance from M larger than $c_0 t$. The influence of these points has no physical basis and introduces distortions into the values of $\partial z/\partial t$ as the cost of filtering out the gravitational waves.

However, to correct this shortcoming of the filtered equations it is not necessary to return fully to the primitive equations. It is sufficient, for example, to replace the left side,

$$\left(\mathfrak{G}\nabla^2 - \frac{1}{L_0^2} \right) \frac{\partial z}{\partial t},$$

by

$$\left(\mathfrak{G}\nabla^2 - \frac{1}{L_0^2} - \frac{1}{c_0^2} \frac{\partial^2}{\partial t^2} \right) \frac{\partial z}{\partial t}$$

in the quasi-geostrophic-approximation equation (7.6) and to use the approximately "consistent" initial values of the velocity and pressure fields for the equation obtained; such a procedure was recommended in my paper.[36]

Incidentally, turning from the filtered equations to the primitive equations (or restoring the hyperbolic operator in the filtered equations) for the purpose of increasing the accuracy of forecasts of synoptic processes so far does not appear inevitable, since the accuracy that can in principle be provided by the filtered equations, particularly the quasi-solenoidal-approximation equations, has in practice not yet been achieved in specific calculations. This is so primarily because of the errors in the numerical calculation. These errors occur in approximating the continuous pressure and velocity fields by their values at a finite number of points on a certain space-time grid and the corresponding replacement of the differential operators that enter into the dynamic equations by difference operators. (Moreover, in the initial data there are errors that arise because of the inaccurate formulation of the boundary conditions on the curved underlying surface.)

10. Vertical Structure of Synoptic Processes

One of the first tasks in integrating the prognostic equations is to describe the *vertical structure* of synoptic processes. This is necessary because the equations contain a second derivative with respect to the vertical coordinate p [which occurs in operator $\mathfrak{F}$ in equation (7.4) and in the invariant F in (8.2)] or an equivalent double integration [operator $\mathfrak{G}$ in (7.6)].

The vertical structure of synoptic processes is simplest in the particular case of the *equivalent-barotropic* atmosphere (see my paper[42]): Deviations $z(x,y,p,t)$ of the altitudes of isobaric surfaces from their standard-atmosphere values $z(p)$ here have the form

$$z(x,y,p,t) = z_0(x,y,t)\psi_0(p). \tag{10.1}$$

The function $z_0(x,y,t)$, which is the single "parameter" of the barotropic model of the atmosphere, can be taken as the altitude of the isobaric surface at some middle level in the troposphere (about 500 mb). In the general case of the baroclinic atmosphere, the function $z(x,y,p,t)$ can be approximated by the expression

$$z(x,y,p,t) = \sum_{n=0}^{N-1} z_n(x,y,t)\psi_n(p), \tag{10.2}$$

where $\psi_n(p)$ are certain fixed functions, and $z_n(x,y,t)$ are parameters that can always be expressed in terms of the values of $z(x,y,p,t)$ at the given levels $p = p_n$ (so that multiparameter models of vertical structure of synoptic processes are equivalent to multilevel models). For no finite number N can expression (10.2) serve as an exact solution of prognostic equations. However, as an approximation these equations can be replaced by the corresponding equations for the parameters of the given model. In actual numerical forecasts, models with two or three parameters have been employed; and in experiments, there have been models with an even greater number of parameters (or levels). In addition, the functions $\psi_n(p)$ have been assigned by proceeding either from a set of qualitative assumptions concerning the vertical structure of synoptic processes or from a consideration of the convenience of interpolating between given levels. But here, naturally, arises the question of the optimum choice of these functions.

With the aid of statistical considerations, one can introduce an optimization criterion. This is done by considering the values of $z(x,y,p,t)$ for fixed t at different points (x,y) to be particular expressions of some random function $\psi(p)$ that is characterized by the correlation function $\mathfrak{B}(p_1,p_2) = \overline{\psi(p_1)\psi(p_2)}$ [the bar signifies a mathematical expectation or average with respect to (x,y); without loss of generality the average value $\overline{x(p)}$ is here assumed to be zero]. We can approximate the function $\psi(p)$ by the sum of the first N terms of its expansion into the complete orthonormal system of the functions $\psi_n(p)$. The mean square error of this approximation is given by

$$\sigma_N^2 = \int \overline{\left| \psi - \sum_{n=0}^{N-1} z_n \psi_n \right|^2} \, dp.$$

From the general theory of random functions, it follows (Obukhov[62,63]) that for any given N, σ^2 will have its minimal value if for $\psi_n(p)$ we choose the characteristic functions of the "dispersion operator" $\mathfrak{B}(p_1, p_2)$ in the integral equation

$$\int \mathfrak{B}(p_1, p_2) \psi(p_2) \, dp_2 = \mu \psi(p_1). \tag{10.3}$$

This choice of functions $\psi_n(p)$ is optimum from a statistical point of view. Under this condition the characteristic values μ of the operator $\mathfrak{B}(p_1, p_2)$ can be taken as the values of the dispersions of the expansion coefficients $z_n = \int \psi \psi_n \, dp$, and pairs of these coefficients are not correlated. Obukhov[63] used such a statistically optimal treatment for describing the vertical structure of the field $\psi(p) = \partial z / \partial t$. In two real examples of such fields, which he examined at a number of discrete levels $p = 1000, 850, 700, 500,$ and 300 mb, the first term of the optimal expansion (corresponding to the barotropic model of the atmosphere), accounted for about 70% of the total dispersion of the field $\partial z / \partial t$; the sum of the first two terms accounted for more than 90%; and the sum of the first three, 97%; the remaining terms represent only a very small fraction of the total dispersion and are therefore roughly approximated. It follows that in models of the vertical structure of synoptic processes, two to three parameters or levels may suffice (a greater number of levels can be useful for a detailed calculation of frictional effects in the planetary boundary layer of the atmosphere and for forecasting the condition of the stratosphere). Note that in Obukhov's two examples the optimum orthonormal functions $\psi_n(p)$ differed very slightly from each other. Special re-

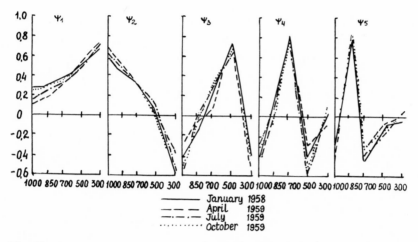

Fig. 13   Eigenfunctions $\psi_n(p)$ of the dispersion operator $\mathfrak{B}(p_1,p_2)$ of the field $\psi(p) = \partial z/\partial t$ at the different seasons of the year (curves 1, 2, 3, 4). After Rukhovets.[64]

search by Rukhovets[64] has demonstrated a small regional and seasonal variation in such functions, both for the field $\psi(p) = \partial z/\partial t$ and for the vertical structure of the fields of the zonal and meridional components of wind velocity and certain other synoptic fields; the first five functions $\psi_n(p)$ for the field $\partial z/\partial t$ at different seasons of the year are given as an example in Fig. 13.

This method of statistically optimal anlysis was applied in earlier studies by Fukuoka,[65] Lorenz,[66] White et al.,[67] and Bagrov[68] to describe the horizontal structure of meteorological fields with a view toward their classification and statistical forecasting. After Obukhov's work[63] this method received very wide application (for example, see Holström's article[69]).

Another method for naturally choosing the functions $\psi_n(p)$ in expansion (10.2) is to use the characteristic functions of the "dynamic operator" $\mathfrak{H}$, which figures in the main linear part of the prognosticative equation (which we shall write in the form $\partial\psi/\partial t = \mathfrak{H}\psi$). For some general condition, these functions $\psi_n(p)$ will coincide with the characteristic functions (mentioned above) of the "statistical operator" $\mathfrak{B} = \overline{\psi(p_1)\,\psi^*(p_2)}$ (here for the sake of generality we allow for complex functions ψ, and the asterisk denotes a complex-conjugate quantity). Indeed, differentiating this expression for $\mathfrak{B}$ with respect to time t, substituting $\partial\psi/\partial t$ for $\mathfrak{H}\psi$, and using the condition $\mathfrak{H}^* = -\mathfrak{H}$ (which guarantees conservation of energy), we obtain

$$\frac{\partial\mathfrak{B}}{\partial t} = \mathfrak{H}\mathfrak{B} - \mathfrak{B}\mathfrak{H}. \qquad (10.4)$$

Thus, under the condition of statistical equilibrium, when the dispersion operator $\mathfrak{B}$ doesn't depend on time t and the left side of equation (10.4) goes to zero, operators $\mathfrak{B}$ and $\mathfrak{H}$ are commutative. Consequently, they have the same characteristic functions (Monin and Obukhov[70]). The connection between these characteristic functions and the "dynamic operator" $\mathfrak{H}$, which doesn't depend on time, can provide an explanation of their statistical stability (the small seasonal and regional variation) that was noted above. This, however, has no bearing on the dispersions μ_n of the coefficients z_n: Within the limits of linear theory their values are in general arbitrary, but in fact they are established as a result of weak nonlinear interactions; therefore their dispersions turn out to be significantly less stable statistically than the characteristic functions of operators $\mathfrak{B}$ and $\mathfrak{H}$. A similar method for choosing the functions $\psi_n(p)$ in expansion (10.2) was proposed in a paper by Gavrilin,[71] where for $\psi_n(p)$ he used the characteristic functions of the "vertical" operator $\mathfrak{H} = (\partial/\partial p)\,(p^2/\alpha_0{}^2)\,(\partial/\partial p)$ (for $\alpha_0{}^2 = \text{const}$), which enters into the three-dimensional elliptical operator $\mathfrak{F}$ of equation (7.5); operator $\mathfrak{F}$ enters into equation (7.4) of the quasi-geostrophic approximation. These characteristic functions were earlier found by White et al.[67] as solutions of the equation $\mathfrak{H}\psi = -\mu\psi$ with boundary conditions $p(\partial\psi/\partial p) + \alpha_0{}^2\psi = 0$ at $p = p_0$ (corresponding to the vanishing of the vertical velocity w at the underlying surface) and $p|\psi|^2 < \infty$ as $p \to 0$ (corresponding to the fact that the kinetic energy is bounded at the upper limit of the atmosphere). Such boundary conditions were used above in obtaining the integral form (7.6) of the equation arising from the quasi-geostrophic approximation. The spectrum of the eigenvalues μ of the operator $\mathfrak{H}$ contains the isolated point $\mu = 1 - \alpha_0^2$ and a straight line for $1/4\alpha_0^2 < \mu < \infty$ (in the limit as $\alpha_0^2 \to 0$, i.e., in going over to a barotropic atmosphere, only a single isolated point remains in the spectrum). The isolated point corresponds to the eigenfunction $\psi_0(p) = (p_0/p)^{\alpha_0^2}$, and the remaining points of the spectrum correspond to the functions

$$\psi_\nu(p) = \sqrt{\frac{p_0}{p}}\left[2\nu \cos\left(\nu \ln\frac{p_0}{p}\right) - (1 - 2\alpha_0^2)\sin\left(\nu \ln\frac{p_0}{p}\right)\right], \quad (10.5)$$

where $\nu = (\alpha_0^2\mu - \frac{1}{4})^{1/2}$. These eigenfunctions are not normalized; aside from this they are very similar to the statistically optimal functions of Fig. 13. Gavrilin[71] considered a bounded layer of the atmosphere $p_0 \geq p \geq p_h$ (with the boundary condition $p(\partial\psi/\partial p) + \alpha_0^2\psi = 0$ at the upper limit of the atmosphere); in this case the continuous part of the spectrum breaks up into a denumerable set of points $\nu = \pi n/\ln (p_0/p_h)$, $n = 1,2,\ldots$, and the form of the eigenfunctions remains unchanged. For the parameters $z_n(x,y,t)$ of expansion (10.2), the following equations are obtained:

$$\mathfrak{F}_n\frac{\partial z_n}{\partial t} = -\frac{g}{l}\left\{\frac{1}{N_n^2}\sum_{p,q} a_{npq}[z_p,\mathfrak{F}_q z_q] + \frac{\partial l}{\partial y}\frac{\partial z_n}{\partial x}\right\}, \quad (10.6)$$

where $\mathfrak{F}_n = (g/l)(\nabla^2 - \mu_n/L_0^2)$ (μ_n are the eigenvalues of the operator $\mathfrak{H}$; $L_0/(\mu_n)^{1/2}$ plays the role of the scale of the horizontal inhomogeneities of the field z_n), N_n are the norms of the eigenfunctions ψ_n, and finally $a_{npq} = \int \psi_n\psi_p\psi_q\,dp$. These equations show that the time variations of each component $z_n\psi_n$ of the field z are determined by the interactions among all such components. Any three components $z_n\psi_n$, $z_p\psi_p$, and $z_q\psi_q$ interact both directly (this direct interaction is described by the coefficients a_{npq}, which are symmetrical with respect to their three indices) and indirectly, through all the remaining components. Such a structure of changes in the pressure field as the result of direct interactions among triads of eigenvalues is a consequence of the quadratic form of the non-linearities in the hydrodynamic equations. In practice it is necessary to consider only a finite number of "parameters" z_n. The equation for z_0 in such a case describes "barotropic" pressure changes. If we put $z_0 \psi_0 = z^*$, then this equation is of the form

$$\mathfrak{F} \frac{\partial z^*}{\partial t} = \frac{g}{l} \left\{ \frac{\psi_0}{N_0^2} \int [z, \mathfrak{F} z] \psi_0 \, dp + \frac{\partial l}{\partial y} \frac{\partial z^*}{\partial x} \right\}. \qquad (10.6')$$

As $\alpha_0^2 \to 0$ it goes over into equation $(7.6')$ for the barotropic model. For the "baroclinic" component $z' = z - z^*$, we obtain the equation

$$\mathfrak{F} \frac{\partial z'}{\partial t} = -\frac{g}{l} \left\{ [z, \mathfrak{F} z] - \frac{\psi_0}{N_0^2} \int [z, \mathfrak{F} z] \psi_0 \, dp + \frac{\partial l}{\partial y} \frac{\partial z'}{\partial x} \right\}. \qquad (10.6'')$$

From the system of the two equations (10.6) and $(10.6')$ it is possible to determine both of the components z^* and z' of the variations in the pressure field and thereby directly estimate the role of baroclinic effects. Presumably these effects will be largest in the frontal zones (between air masses with different properties), and outside these zones the "barotropic" component z^* will represent the main component of the variations in pressure.

11. Difference Schemes
In a way similar to the transition from a continuous vertical coordinate to a discrete set of "parameters" (or levels) that describe the vertical structure of the atmosphere, it is necessary, in the numerical integration of the prognostic equations, to transfer from the continuous horizontal coordinates and time to a discrete set of spatial points and instants of time and to replace the differential operators appearing in the equations with difference operators.

Basically, the important conditions turn out to be those concerning the similarity between the solutions of the difference equations that are obtained and the solution of the original differential equations. In this section we shall deal very briefly with these conditions.

First, for simplicity we shall consider the errors that arise in replacing the derivatives by finite differences along the horizontal coordinates only. The nature of these errors was investigated by Obukhov[72] in the case of the most simple prognostic equation, the one-dimensional transfer equation

$$\frac{\partial \psi}{\partial t} = - U \frac{\partial \psi}{\partial x} \tag{11.1}$$

with a constant transfer rate U. This equation is obtained, for example, from the quasi-geostrophic approximation applied to the barotropic atmosphere (7.6') (ignoring the terms with $1/L_0^2$ and those with $\partial l/\partial y$) in the particular case when the values of z can be represented in the form of a sum $z = -(U/g)y + \psi$, where ψ satisfies the additional condition $\nabla^2 \psi = -k^2 \psi$ (this condition is exactly satisfied for sinusoidal waves). However, equation (11.1) is not simply a model but has wider significance: By using the so-called splitting method (Marchuk[57-60]), it is possible to reduce the problem of numerical integration of the general prognostic equations (including the primitive equations) to solving equations of this form (only with a variable flow speed U).

Let us denote $\psi(x_n, t)$ by $\psi_n(t)$. Suppose we transform from continuous horizontal coordinates to a discrete set of points $x_n = n\Delta x$ $(n = \ldots, -1, 0, 1, \ldots)$ and replace the derivative $\partial \psi/\partial x$ at the points x_n by $(\psi_{n+1} - \psi_{n-1})/2\Delta x$. Then, in place of the partial differential equation (11.1), we obtain an infinite system of ordinary differential equations relating the functions $\psi_n(t)$ of the form

$$\frac{d\psi_n}{dt} = - U \frac{\psi_{n+1} - \psi_{n-1}}{2\Delta x}. \tag{11.2}$$

The solution of the original equation (11.1) for an arbitrary initial

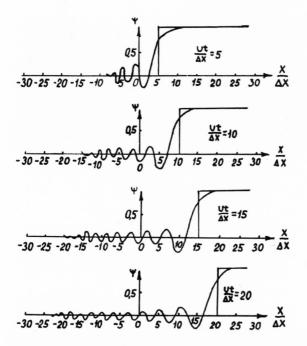

Fig. 14 Solution to the differential-difference transfer equation (11.2), corresponding to a model of a moving front. After Obukhov.[72]

condition $\psi(x,0) = \Psi(x)$ is of the form $\psi(x,t) = \Psi(x - Ut)$. This corresponds to the solution of equations (11.2) with the initial condition $\psi_n(0) = \Psi_n = \Psi(x_n)$, which is of the form

$$\psi_n(t) = \sum_m \Psi_{n-m} J_m\left(\frac{Ut}{\Delta x}\right),\tag{11.3}$$

where J_m is the Bessel function. Let us examine, in particular, the case

$$\Psi(x) = E(x) = \begin{cases} 0 & \text{for } x < 0, \\ 1 & \text{for } x \geq 0, \end{cases}$$

which is a model of an atmospheric front. The solution of the initial equation (11.1) here will be a shifting front $\psi(x,t) = E(x - Ut)$.

On the other hand, the solution of equations (11.2) (see Fig. 14 for graphs of the solution at $Ut/\Delta x = 5, 10, 15$, and 20) is a "frontal zone" that is gradually diffused with time. From behind this zone, parasitic waves spread out "upstream." These appear because of the jump (the sharp change over the distance Δx) in the initial values of Ψ. This is the nature of the errors that arise in replacing equation (11.1) by (11.2).

One means of dealing with the parasitic waves (while increasing the time-diffusion of the frontal zone, however) can be to use a spatial smoothing, which is realized, for example, by adding the term $\alpha(\psi_{n+1} - 2\psi_n + \psi_{n-1})$ to the right side of equation (11.2); the quantity $\alpha(\Delta x)^2$ is then analogous to the diffusion coefficient. In particular, for $\alpha = |U|/2\Delta x$ in the case $U > 0$, equation (11.2) is replaced by the equation

$$\frac{d\psi_n}{dt} = -U\frac{\psi_n - \psi_{n-1}}{\Delta x}; \qquad (11.2')$$

the solution of this equation, which is of the form

$$\psi_n(t) = \sum_{m=0}^{n} \Psi_{n-m}\frac{(Ut/\Delta x)^m}{m!}e^{-Ut/\Delta x}, \qquad (11.3')$$

does not contain any perturbations spreading "upstream." Thus, replacing the "centered" difference $\psi_{n+1} - \psi_{n-1}$ by the "directed" difference $\psi_n - \psi_{n-1}$ turns out to be equivalent to a certain kind of smoothing and excludes parasitic waves from the solution.

Let us now consider some more complete difference schemes that are used in practice in the numerical integration of the prognostic equations, namely, the schemes in which, in the initial equations, not only the derivatives with respect to the spatial coordinates but also those with respect to time are replaced by the finite differences. As above, we shall use the transfer equation (11.1) for illustration.

Transforming from continuous time to a discrete set of time instants $t_m = m\Delta t$ $(m = 0, 1, 2, ...)$, denoting $\psi_n(t_m)$ by ψ_n^m, using, let us say, equation (11.2') and in it replacing the derivative $d\psi_n/dt$ at instant t_m, for example, by $(\psi_n^{m+1} - \psi_n^m)/\Delta t$, we obtain instead of the partial differential equation (11.1) the difference equation

$$\frac{\psi_n^{m+1} - \psi_n^m}{\Delta t} = -U\frac{\psi_n^m - \psi_{n-1}^m}{\Delta x}, \qquad (11.4)$$

which has to be solved for given initial conditions $\psi_n^0 = \Psi_n$. In considering the nearness of the solution ψ_n^m of the difference equation (11.4) to the solution $\psi(x,t)$ of the original differential equation (11.1), we shall assume the possibility of a limiting transition $\Delta x \to 0$ and $\Delta t \to 0$, assuming, for example, that the ratio $\Delta x/\Delta t$ always remains constant. Under these circumstances, the following three concepts become essential (they are explained in detail in the book by Godunov and Ryaben'ky,[73] for example):

1. The *order of approximation* of the differential equation by the difference equation for some solution $\psi(x,t)$; that is, the order of smallness of the standard error of approximation (relative to Δx) on the space-time grid (x_n, t_m) that is being considered.

2. The *convergence* of the solution ψ_n^m of the difference equation to the solution $\psi(x,t)$ of the differential equation as $\Delta x \to 0$.

3. The *stability* of the difference scheme. This can be qualitatively described as the condition that the difference equation has no solutions increasing rapidly with decreasing Δx.

The condition of stability is important, because if the difference equation is stable and it approximates the differential equation at some solution $\psi(x,t)$, then the solution of the difference equation will converge to $\psi(x,t)$ for $\Delta x \to 0$.

It is possible to show that, for equation (11.4) with $U > 0$, the stability criterion has the form

$$\frac{\Delta x}{\Delta t} \geq U; \qquad (11.5)$$

and if $\Delta x/\Delta t < U$, equation (11.4) will be unstable. This criterion indicates that for a given spatial interval Δx, it is necessary to choose a sufficiently small time interval $\Delta t \leq \Delta x/U$ in order to guarantee computational stability; for example, if for numerical weather forecasts a spatial grid with an interval $\Delta x = 300$ km is used, and if we put $U \sim 10$ m/sec, then a time interval $\Delta t \leq 8\frac{1}{3}$ h must be chosen.

The stability criterion, sometimes called the "criterion of linear stability," was established for a wide class of difference schemes by Courant, Friedrichs, and Lewy in 1928.[74] In order to make its meaning clear, consider, for example, the value of ψ_0^{m+1} and note that according to (11.4) it is expressed in terms of the initial values of Ψ_n with subscripts $n = 0, -1, -2, \ldots, -(m+1)$. Or, if we denote $(m+1)\Delta t$ by t, then the value of $\psi(0, t)$ depends on the initial values $\Psi(x)$ at points lying in the interval $-(\Delta x/\Delta t)t \leq x \leq 0$. At the same time, for the solution $\psi(x, t)$ of the original equation (11.1) the value of $\psi(0, t)$ is determined by the initial value of $\Psi(x)$ at the point $x = -Ut$. If criterion (11.5) is not satisfied, then the point $x = -Ut$ will lie outside the interval $-(\Delta x/\Delta t)t \leq x \leq 0$; and by changing the value of $\Psi(x)$ at this point but preserving the values of $\Psi(x)$ in the indicated interval, we can disturb the convergence of ψ_n^m to $\psi(x, t)$, so that equation (11.4) will be unstable. This reasoning can apparently be generalized: For convergence, it is always necessary that the region in which initial data are assigned for computing the space-time solution of the *difference* equation at a fixed point M contain, for sufficiently small Δx, all points of the region in which initial data are assigned for computing the solution of the original *differential* equation at the same point M.

A type of computational instability that is quite different from that which arises in violating the CFL condition [that is, the Courant-Friedrichs-Lewy condition (11.5)] can occur as a result of the *nonlinearity* of the prognostic equations. The nonlinear terms of the dynamic equations describe the interactions between components of motion with different scales. These interactions lead, in particular, to the production of small-scale components of motion (usually in the form of small-scale turbulences) due to the decay of the large-scale components. The interactions thereby lead to an energy transfer from large-scale to small-scale components or, figuratively speaking, to the appearance of an energy flow along the scale spectrum in the direction from large to small scales (this "spectral energy transfer" was already mentioned in section 3 in the description of "synoptic oscillation" in meteorological elements). The energy transferred along the scale spectrum finally ends up in the region of smallest scales, in which motions are damped through the action of viscous forces—i.e., in which frictional energy dissipation occurs.

But in the numerical solution of the prognostic equations, the continuous space is replaced by a discrete set of points separated by distances Δx, so that Δx plays the role of the smallest scale of motion described by the prognostic difference equations. In practice, this scale Δx is given a value of several hundred kilometers, so that it greatly exceeds the scales of the region of frictional energy dissipation. For this reason, if we do not take energy dissipation into account (furthermore, recall that the adiabatic approximation is generally employed in the theory of short-range weather forecasting, i.e., neither external energy generation nor dissipation of energy is taken into account), we find that the energy transferred along the scale spectrum ultimately reaches the smallest scale of the order Δx, where it *accumulates* without dissipating, so that the inhomogeneities of synoptic fields with scales on the order of Δx (of the spatial grid that is used) will gradually increase with time. This increase of small-scale inhomogeneities is called nonlinear instability.

Nonlinear instability not only can occur but in fact does occur in the numerical integration of equations that model the synoptic processes; see, for example, Phillips's paper,[75] in which he analyzes the occurrence of nonlinear instability in the simple barotropic equation of vorticity transfer (7.6′) (which was subjected to some additional simplifications). In order to deal with this nonlinear instability, it has been suggested that a term describing the large-scale horizontal diffusion of momentum and heating be introduced into the hydrodynamic equations (Smagorinsky,[53, 76] Leith[77]). Or one could provide that the basic adiabatic invariants enumerated in section 4 be preserved in the finite-difference scheme (Mintz,[78] Lilly[79]). Other methods have also been proposed (Schuman,[80] Platzman[81]). A survey of the finite-difference schemes employed in solving the primitive equations can be found, for example, in the article by Fisher.[82] *

12. Weather Prediction

In the quasi-geostrophic approximation we deal with equations (7.4) or (7.6), which describe the synoptic oscillations of the field of the altitudes of the isobaric surfaces $z(x,y,p,t)$, i.e., the *pressure field* in the atmosphere [in the quasi-solenoidal approximation, in equations (8.2) and (8.3), the field of the stream function ψ is added to the field z]. This naturally raises the question of the extent to which it is possible, from the predicted changes in the atmospheric pressure field (which in themselves can hardly be sensed by humans), to make predictions concerning those features of the weather to which people are very sensitive. These are, primarily, the air temperature, the wind, and, finally, the variable cloudiness and precipitation, which give the weather on our planet its peculiar charm. To this question, it is possible to give a fairly optimistic reply.

12.1. In fact, knowledge of the field $z(x,y,p,t)$ permits one, in the first place, to calculate the *air temperature* T according to the equation $T = -(g/R)(\partial z/\partial \ln p)$ (introduced in the small print on page 33).

---

*Other studies dealing with the development of the methods for the integration of primitive equations include: A. Arakawa, *J. Comput. Phys.* **1**, 119–143 (1966), Y. Kurihara, *Monthly Weather Rev.* **93**, 399–415 (1965); Y. Kurihara and J. L. Holloway, Jr., *Monthly Weather Rev.* **95**, 509–530 (1967); D. L. Williamson, *Tellus* **20**, 642–653 (1968); R. Sadourny, A. Arakawa, and Y. Mintz, *Monthly Weather Rev.* **96**, 351–356 (1968); F. Baer and R. L. King, *J. Comput. Phys.* **2**, 32–60 (1967).

The accuracy of the temperature forecast, moreover, will be limited by the accuracy of the calculation of the derivative $\partial z / \partial \ln p$ [i.e., by the accuracy of the description of the vertical structure of the field z (see section 10)], as well as by the accuracy of the prediction of the field z itself. In estimating the second of these factors, it is essential that only the *synoptic* oscillations of the field z are being predicted. These have rather large spatial and temporal scales. It is also important that they are only being forecast in the *adiabatic* approximation. For this reason, from the predicted field z it is only possible to calculate the smoothed temperature field, i.e., the "synoptic" temperature background. In nature, small-scale oscillations and the diurnal variation produced by nonadiabatic factors (for example, in the boundary layer at the underlying surface) are superimposed on this background. The forecasting of the daily course of the surface temperature can be carried out separately, as is done in current practice.

12.2. Second, from the field z it is possible to calculate the horizontal components u and v of the *wind field* at any level by using the quasi-geostrophic approximation formulas (7.1) (in the quasi-solenoidal approximation u and v are determined by the stream function ψ, which is calculated simultaneously with z). Knowing the speed and direction of the wind at various altitudes is very important, for example, in aviation.

One should bear in mind that only the solenoidal part of the horizontal wind field is given by the above-mentioned equations (and inexactly at that—with a relative error on the order of Ki in the quasi-geostrophic approximation). The solenoidal part can be identified with the entire wind field only as an approximation (again, with a relative error of the order Ki in the quasi-geostrophic approximation); it is in no way possible to estimate from it the divergence of the wind velocity D (in this approximation it is simply equal to zero).

12.3. Third, from the field z it is possible to calculate the field of $w^* = dp/dt$, which replaces the *vertical velocity* in (x,y,p) coordinates (from w^* the divergence $D = -\partial w^*/\partial p$ can be determined). That is to say, a diagnostic equation (i.e., one containing no time derivatives) can be found for w^*. This is done by differentiating the balance equation (8.3) with respect to p and t and then eliminating the derivative $\partial \psi / \partial t$ from it by using the equation for the transfer of the potential vorticity (8.2) (though here it is more convenient to use the vorticity equation that is written with the same degree of accuracy in the form $\partial \nabla^2 \psi / \partial t + [\psi, \nabla^2 \psi + l] = l \, \partial w^* / \partial p)$ and by eliminating the derivative $\partial z / \partial t$ by using the adiabaticity equation (7.2). Then for w^* we obtain an elliptic equation whose right side is

$$\frac{\alpha_0^2 c_0^2}{p^2} \nabla^2 w^* + \mathfrak{U} l \frac{\partial^2 w^*}{\partial p^2} = - g \nabla^2 \left[\psi, \frac{\partial z}{\partial p}\right] + \mathfrak{U} \frac{\partial}{\partial p} [\psi, \nabla^2 \psi + l],$$

(12.1)

where $\mathfrak{U}$ is the operator $(\nabla \cdot l\nabla) \, (\nabla^2)^{-1}$ (here we have left out other small terms that come from the term $2[\partial \psi / \partial x, \partial \psi / \partial y]$ in the balance equation). In the quasi-geostrophic approximation it is sufficient to replace the stream function by gz/l in (12.1). In this approximation (and for $l \approx$ const, when $\mathfrak{U} \equiv l$), Knighting,[83] for example, gave instances of the construction of the field w^* from z by solving equation (12.1). He used real data for the values of the field z at nine levels: $p = 1000, 900, 800, \dots, 200$ mb. His grid had 480 points with spacings of about 160 miles, and it covered a considerable portion of the northern Atlantic and western Europe. He solved the finite-difference analog of equation (12.1) under zero boundary conditions for w^* on the boundaries of the region considered (including the lower limit $p = 1000$ mb and the upper limit $p = 200$ mb).

One of Knighting's examples (for 01[h] of 2 December 1958) is shown in Fig. 15, where the profile of $w^*(p)$ (in mb/h) is given at each of the 16×12 internal points of the horizontal grid (negative w^*, corresponding to updrafts, are plotted on the right; and positive w^*, corresponding to downdrafts, are plotted to the left of the origin). The solid lines in the illustration show the isolines of the altitudes z (in meters) of the isobaric surface $p = 1000$ mb, and the dotted lines show the atmospheric fronts. This example and others show that the profiles of $w^*(p)$ calculated from equation (12.1) vary quite smoothly and regularly from point to point. The profiles also correspond closely to the customary notions of synopticians, for example, concerning the rising of air ahead of pressure troughs and the

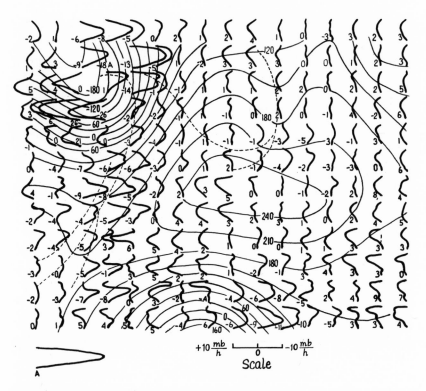

Fig. 15 Profiles of $w^*(p)$ in the layer 1000–200 mb and the isolines of equal altitudes of the isobaric surface 1000 mb above the North Atlantic and western Europe for 01^h 2 December 1958. After Knighting.[83]

descent of air behind them. If the values of w^* are large, then their maximum (and therefore also the zero value of the divergence $D = -\partial w^*/\partial p$) is reached at a middle level in the troposphere, near 500–600 mb; on the average, the profiles of $w^*(p)$ have an approximately parabolic shape.

However, we must remember that the vertical velocity field w [or $w^* = (pg/RT)(d_h z/dt - w)$] is much more sensitive to factors other than the pressure field and the horizontal wind field that we have not yet taken into account. It is sensitive primarily to the variations in the earth's surface and to nonadiabatic factors. The former can be taken into account by specifying the shape of the earth's surface by the function $z = \zeta(x,y)$, so that for smooth flow over the relief, we get $w = u(\partial\zeta/\partial x) + v(\partial\zeta/\partial y) \approx [\psi,\zeta]$ (this condition must be written for $z = \zeta$, but for small ζ it can be changed to the level $z = 0$ or $p = p_0$ as an approximation).

The most important of the nonadiabatic factors is the surface friction, which is characterized by the horizontal vector τ of the frictional stress against the earth's surface. When this friction is taken into account, the value of w at the upper boundary of the friction layer (or again, at $p = p_0$ for an approximation) can be determined by the equation $w = (1/l\rho)$ curl τ. In describing the friction layer by the so-called Eckman model, with an altitude-independent coefficient of turbulent viscosity K, we get

$$w = \frac{g}{l}\sqrt{\frac{K}{2l}}\nabla^2 z.$$

Both of the effects indicated—the presence of the relief and the surface friction—must be taken into account in the boundary condition for equation (12.1). This boundary condition is written in the form

$$w^* = \frac{p_0 g}{RT_0}\left(\frac{d_h z}{dt} - [\psi,\zeta] - \frac{g}{l}\sqrt{\frac{K}{2l}}\nabla^2 z\right) \qquad \text{at } p = p_0. \tag{12.2}$$

With the aid of (12.1) and (12.2) we can then calculate the field w^* from the given fields z and ψ. In a more complete formulation of the problem, it is necessary to take into account the influence of the relief and the friction on the fields z and ψ. This is done by using the boundary condition (12.2) in solving the equation of the quasi-geostrophic approximation (7.4) [this introduces additional terms in the right side of equation (7.6)] or the equations of the quasi-solenoidal approximation (8.2), (8.3). However, here it is natural to also take the remaining nonadiabatic effects into account; of these, the release of the latent heat of condensation in clouds is especially important for the field w^*.

12.4. Knowledge of the field ψ $(gz/l$ in the quasi-geostrophic approximation) and the field w^* makes it possible to calculate the displacements of various substances in the atmosphere. In the case of conservative substances, for this purpose it is possible to use the transport equation in the form

$$\frac{dq}{dt} = D\{q\}, \tag{12.3}$$

where q is the specific concentration of the substance (i.e., the ratio of its mass in a unit volume of air to the total mass of air and substance contained in that volume), with

$$\frac{dq}{dt} = \frac{\partial q}{\partial t} + [\psi, q] + w^* \frac{\partial q}{\partial p}, \tag{12.3'}$$

$$D\{q\} = -\frac{1}{\rho} \operatorname{div} \mathbf{Q},$$

where $\mathbf{Q}$ is the density of the diffusion flux of the impurity. For the most part, this flux is produced by turbulent diffusion, which is assumed to be a linear function of the gradient ∇q of the field q. However, in the free atmosphere (i.e., above the planetary boundary layer of the atmosphere) during time intervals that are not too large, the turbulent diffusion of substances [i.e., the right side of equation (12.3)] is frequently neglected.

If the water vapor in the air is in an unsaturated state, then it is a conservative substance. Equation (12.3) is therefore suitable to describe the evolution of the field of the specific humidity q (in this case, by the way, knowledge of the magnitude of w^* makes it possible to estimate the adiabatic cooling of rising air particles and the heating of falling air particles by means of the simple formula $dT/dt = [(\kappa - 1)/\kappa]\,(T/p)w^*)$. Sometimes, instead of using q it is more convenient to describe the moisture of the air by the dew point T_m. The dew point is the temperature at which the air, with fixed relative humidity q and fixed pressure p, becomes saturated (over a plane surface of water). The dew point is determined from the relation $q = (R/R_v)\,(e_m(T_m)/p)$, where R and R_v are the gas constants of dry air and water vapor, and $e = e_m(T)$ is the partial pressure of the saturated water vapor. Substituting this equation for q in equation (12.3) (neglecting its right-hand side) and using the Clausius-Clapeyron equation (4.4) for $e_m(T)$, we obtain for the temperature dew point depression $\Delta = T - T_m$ the equation

$$\frac{d\Delta}{dt} = \frac{\kappa - 1}{\kappa}\frac{T}{p}\left(1 - \frac{\kappa}{\kappa - 1}\frac{R_v\,T_m^{\,2}}{\mathscr{L}\,T}\right)w^*. \tag{12.4}$$

12.5. Lewis[84] proposed a very simple method for predicting *the amount of cloudiness and the presence of precipitation.* It was based on the use of the empirical relation of these phenomena to the values of w^* and Δ, which can be calculated with the aid of equations (12.1) and (12.4). This empirical connection is shown in Fig. 16, where the value of Δ at the level $p = 700\,\mathrm{mb}$ is the abscissa, and the ordinate is the value of w in the middle troposphere (approximately $w^* \approx -\rho g w$); the different symbols on the nomogram represent weather phenomena actually observed (during the period 15–28 March 1960, in Japan, according to Nabeshima[85]). Similar methods of forecasting the cloudiness and precipitation were developed or used by Shvets,[86] Dushkin, Lomonosov, and Lunin,[87] Vederman,[88] Ovsyannikov,[89,90] Kuznetsov,[91] Pedersen,[92] Uspensky,[93] Bagrov,[94] and several other authors.

In cases when a point (Δ, w^*) falls in the precipitation region on Lewis's nomogram, Vederman, following Lamb[95] and Wagner,[96] predicted rain for $z_{500} - x_{1000} \geq f(M)$, and snow for $z_{500} - z_{1000} < f(M)$, where $f(M)$ is the critical thickness of the air layer between the levels 500 mb and 1000 mb, determined for the territory of the U.S.A. Moreover, following Smagorinsky and Collins,[97] Vederman predicted the amount of precipitation in the time δt from the equation $A = (w^*/\rho_w g)\,(r_{500} - r_{1000})\,\delta t$, where ρ_w is the density of liquid water, and $r = q/(1 - q)$ is the mixing ratio (i.e., the ratio of the mass of water vapor in a unit volume to the mass of the dry air in that volume). As it turned out, in cases where precipitation occurs, calculating w^* in the adiabatic approximation with the aid of equation (12.1) results in an underestimation of the values of w^* and therefore also of the amount of precipitation A (by an average of approximately a factor of three). This is readily explained: The heat released in clouds during the condensation of water vapor is not taken into account (Smagorinsky[98]). Following Smebye,[99] Vederman, in calculating the second approximation for w^*, took into account the rate ε of heat release during a time δt by a quantity of precipitation A, obtained with the aid of the first approximation [the parameter ε will enter into the right side of equation (12.1) for w^* as a term $-(R/p)(\nabla^2 \varepsilon/c_p)$].

Note that, strictly speaking, the vertical velocity must determine not the amount of cloudiness but the rate of its change with time. This is especially clear for the first stages of the growth of clouds (on the contrary, precipitation, which is the result of cloud alterations, is perhaps more closely related to w^*). Therefore, Smagorinsky[100] undertook an attempt to calculate the amount of clouds without considering w^* at all by using only the values of the relative humidity. He used the average relative humidity for the cloud layer; for clouds of the lower, middle, and upper levels he chose the layers 1000–800 mb, 800–550 mb, and 550–300 mb. According to Smagorinsky's empirical nomogram, the amount of cloudiness at each level turned out to be approximately a linear function of the corresponding relative humidity averaged over the layer.

Methods of predicting the cloudiness and precipitation from their empirical relationship to the values of Δ and w^* or to the values of the relative humidity at some layer in the troposphere are, of course, very crude. Thus, according to Antonov,[101] the probabilities of precipitation, total cloudiness, or clear weather when the point (Δ, w^*) falls within the corresponding regions on a nomogram (such as the one in Fig. 16) amount to only 70–80%, and for partial cloudiness, only about 50%. Evidently it is necessary to develop other forecasting methods that would start from a concrete description of the physical processes of cloud and precipitation production. Water vapor can then no longer be considered a conservative and passive substance, as was done in the absence of clouds when using equation (12.2); when clouds appear, it is necessary to take into account the phase transitions of the moisture, the nonadiabatic effects of the release or absorption of the latent heat of phase transitions, and also the loss of moisture by precipitation.

One such scheme was developed in a series of papers by Matveyev[102-106] (see also chapter 21, section 3, of his book[107] and

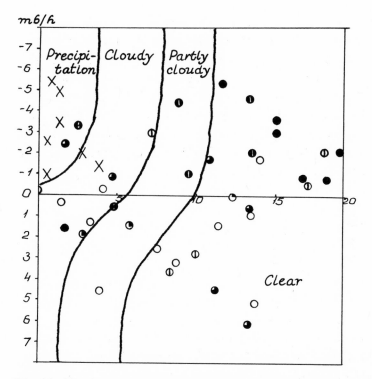

Fig. 16 The amount of large-scale cloudiness and precipitation as empirically related to the values of the dew-point deficit Δ at a level of 700 mb and the rate w of large-scale vertical motions in the middle troposphere, after Lewis;[84] actually observed weather phenomena, after Nabeshima[85] (the blackened fraction of a circle is the cloudiness index, and a cross indicates precipitation).

the following papers: Lushev and Matveyev[108] for examples of predictions of large-scale cloudiness; Bykova and Matveyev[109] for a description of the results of a numerical experiment on the evolution of cloudiness in a cyclone; and Feygel'son and Frolova[110] for several methodological improvements). In this scheme, two phases of moisture in clouds are considered—water vapor with a specific concentration q and water droplets (plus ice) with a specific concentration q_w (called the *specific water content*). For them, the following equations are assumed:

$$\left.\begin{aligned}
\frac{dq}{dt} &= D\{q\} - m, \\[2ex]
\frac{dq_w}{dt} &= D\{q_w\} + m - n,
\end{aligned}\right\} \tag{12.5}$$

where m is the specific rate of condensation (plus sublimation) of water vapor (i.e., the mass of water vapor that condenses or sublimates per unit mass of air in a unit of time), and $n = (1/\rho)\,(\partial Q_w/\partial z)$ is the *rate of precipitation*, where Q_w is the mass flux density of the droplets and crystals of ice produced by their gravitational settling. The latter can be represented in the form $Q_w = -\rho q_w \tilde{W}$, where $\tilde{W}$ is the average rate of precipitation, weighted with weights $r^3 f(r)$. The function $f(r)$ is the probability density of cloud elements of radii r; in this scheme, it is assumed to be known, for example, logarithmically normally distributed for parameters that are related to q_w by some empirical equations (in the paper by Feygel'son and Frolova[110] it is suggested that the rate of precipitation n be determined by the more simple empirical equation $n \sim q_w - q_w{}^*$, where $q_w{}^*$ is the empirically established maximum specific water content of clouds). To equations (12.5) is added the equation of the nonadiabatic heating, which here is conveniently written in the form

$$\frac{dT}{dt} - \frac{\kappa - 1}{\kappa} \frac{T}{p} w^* = D\{T\} + \frac{\varepsilon}{c_p}, \tag{12.6}$$

where $\varepsilon = \varepsilon_r + \varepsilon_q$ is the rate of heat influx (per unit of mass) produced by the radiative heat exchange (ε_r) and the phase transitions of the moisture (ε_q), and $D\{T\} = (1/c_p\rho)$ div $\mathbf{Q}_T$, where $\mathbf{Q}_T$ is the density of the turbulent heat flux.

According to Zilitinkevich and Laykhtman,[111] the vertical component of the turbulent flux of the water vapor in clouds can be written in the form $Q_z = -\rho K(\partial q/\partial z + \beta)$, where $\beta = (c_p/\mathscr{L})(\gamma_a - \gamma_b)$ is the equilibrium gradient of the humidity, K is the coefficient of turbulent diffusion, $\gamma_a = [(\kappa - 1)/\kappa](g/R)$ is the adiabatic temperature gradient, and γ_b is the so-called wet-adiabatic temperature gradient, defined by the formula

$$\gamma_b = \gamma_a \frac{1 + (\mathscr{L}/R)(q_m/T)}{1 + (\mathscr{L}/c_p)(\partial q_m/\partial T)}. \tag{12.7}$$

Introducing β into Q_z is equivalent to removing the term $m_1 = -(1/\rho)(\partial(\rho K\beta)/\partial z)$ from m. The vertical component of the turbulent heat flux here should be written in the form

$$Q_{Tz} = -c_p\rho K\left(\frac{\partial T}{\partial z} + \gamma_b\right) = -c_p\rho K\left(\frac{\partial T}{\partial z} + \gamma_a - \frac{\mathscr{L}}{c_p}\beta\right).$$

The heating ε_q in the atmosphere differs from zero only in clouds where phase changes in the moisture take place; there it is equal to $\mathscr{L}m$, where $\mathscr{L}$ is the latent heat of vaporization (or the latent heat of sublimation, which is close to it in magnitude).

Matveyev's scheme is based on the fact that the equations for the "equivalent temperature" $\Pi = T + (\mathscr{L}/c_p)q$ and the "total specific moisture content" $\tilde{q} = q + q_w$, derived from (12.5) and (12.6) (with certain simplifications), do not contain m and have the same form both in clouds and outside clouds. Having found Π and $\tilde{q}$ from these equations, from the relation $T + (\mathscr{L}/c_p)(R/R_v)(e_m(T)/p) = \Pi$ it is possible to find T and to define clouds as regions in which the difference $\tilde{q} - (R/R_v)(e_m(T)/p)$ is positive; this quantity is then equal to the specific humidity q_w. To avoid an unreliable numerical calculation of this difference, it is suggested in the article by Feygel'son and Frolova[110] that q_w be found from the second equation of (12.5), in which m is defined by the equation

$$
m = \left(1 + \frac{\mathcal{L}}{c_p} \frac{\partial q_m}{\partial T} \right)^{-1} \left[w* \frac{q_m}{p} \left(1 - \frac{\kappa - 1}{\kappa} \frac{T}{q_m} \frac{\partial q_m}{\partial T} \right) \right.
$$
$$
\left. + D\{q_m\} - \frac{\partial q_m}{\partial T} D\{T\} - \frac{\varepsilon_r}{c_p} \frac{\partial q_m}{\partial T} \right], \qquad (12.8)
$$

which is obtained after eliminating dT/dt with the aid of (12.6) from the first equation of (12.5), written for a cloud, i.e., for $q = q_m(T,p)$. Shvets[112] was apparently the first to derive an equation of this type for m. Lebedev[113, 114] used a similar equation in constructing concrete physicomathematical models of clouds.

A more detailed scheme is proposed in the article by Marchuk[115] (see also his book[60]). Besides q and q_w, Marchuk separately introduces the specific concentration of ice crystals q_π. The phase transitions of the moisture are described by the terms $\Sigma_i \alpha_{ij} q_i$ in the expressions for dq_j/dt (the coefficients α_{ij}, satisfying the condition $\Sigma_j \alpha_{ij} = 0$, are given by certain semiempirical equations). Precipitation in the liquid and solid phases is described by very simple empirical equations (which were later used in the paper by Feygel'son and Frolova[110]).

Still more detailed schemes will include the calculation of the characteristics of the microstructure of a cloud. Foremost is the previously mentioned probability density $f(r)$ for cloud elements of radii r (generally speaking, it will depend on the spatial coordinates and the time). It will therefore be necessary to create a synthesis of the dynamics and microphysics of clouds.

The microphysics of clouds has been actively developed in the postwar years, mostly in connection with experiments on the artificial seeding of clouds (for example, by seeding them with crystals of dry ice, silver iodide, or other coagulants), and it has made considerable progress (a survey of which can be found in Fletcher's monograph,[116] for example). If experiments in causing rain still do not yield reliable results, the dispersal of certain kinds of clouds (although temporary) has succeeded with good reliability (failure, by large airports, for example, to use the methods that have been developed is apparently just an indication of organizational inertia). At the same time, the microphysics of clouds has not been sufficiently oriented toward the problem of weather forecasting, and as yet only very little research has been done in this direction. As examples we can cite the papers by Buykov[117] and Shulepov and Buykov.[118, 119] In them, it is assumed that $m = 4\pi N\tilde{r}\chi(q - q_m)$, where $N\tilde{r}$ is the average number of cloud elements in a unit of volume, $\tilde{r}$ is their average radius, and χ is the diffusion coefficient of water vapor. For the probability density $f(r)$ the kinetic equation was used:

$$\frac{df}{dt} - ar^2 \frac{\partial f}{\partial z} + \chi \frac{q - q_m}{q_w} \frac{\partial}{\partial r} \frac{f}{r} = D\{f\}, \qquad (12.9)$$

where ar^2 is the descent velocity of a cloud droplet in still air.

12.6. The synthesis of the dynamics and microphysics of clouds is still a problem for the future. Another problem that is still unsolved is the prediction of *convective clouds and the precipitation falling from them*: Individual convective clouds are in their spatial scale not synoptic but mesometeorological phenomena, so that the theory of sections 7 and 8 is not applicable for their description. At the same time, the amount of precipitation from convective clouds is comparable to the amount of precipitation from the large-scale systems and must be taken into account in describing the synoptic processes. At this time, attempts are being made to find empirical relationships between convective cloudiness and precipitation with a large-scale synoptic background; it is possible that the establishment of such relationships will be aided by the physical and mathematical models of convective cells (of the type constructed by Lebedev[114]).

In searching for these relationships, it will obviously be necessary to distinguish between *free* cumulus convection and *forced* cumulus convection. The former is observed, for example, in middle latitudes over *areas* occupied by cold air masses ("air mass" showers and thunderstorms, determined primarily by the humidity field and the energy of the instability of the lower troposphere, but probably related weakly to the large-scale vertical velocity w^*, which, by the way, corresponds to the settling of air in the anticyclone regions typical of summer convection). Forced cumulus convection is observed on the *lines* of horizontal convergence (on the intertropical convergence zone and on cold fronts in the middle latitudes) and *at points* of convergence (tropical cyclones); it is determined primarily by the humidity field and the magnitude of the convergence D (or the corresponding vertical velocity w^*). According to Charney's idea, cumulus convection itself produces to some degree the horizontal convergence that induces it, so that here there is a bilateral interaction between large-scale and mesoscale processes.

The theory of sections 7 and 8 is based on the adiabatic approximation and the "filtered" equations. It is inadequate not only for the prediction of such mesometeorological phenomena as convective clouds but also for forecasting phenomena with scales on the borderline between the mesometeorological and synoptic regions, such as *tropical cyclones*—hurricanes or typhoons. Tropical cyclones are the most violent weather phenomena on earth (see, for example, the article by Riehl[120] and the books by Riehl[121] and Tiron[122]). Thus, at their centers the lowest air pressure at sea level is observed (see the example in Fig. 17): The record, registered three times, is 890 mb where 1013 mb is normal; moreover, the rate of pressure drop sometimes reaches 40 mb/20 min, and the pressure differences in the horizontal direction frequently exceed 50 mb/50 miles—in which case the wind velocity reaches 100 m/sec. At the center of a hurricane one observes the "eye of the storm," a region of calm (except, of course, for ocean waves) and the partial or complete clearing of the sky (and consequently, descending air motions), with a diameter of about 30 km on the average. The trajectory of hurricanes in the

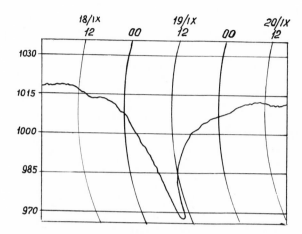

Fig. 17 A barogram of the passing of a hurricane (for 18–20 September 1947, in New Orleans). After Riehl.[120]

Northern Hemisphere is almost always directed from the equator toward the middle latitudes, with many hurricanes first moving toward the northwest and then veering to the northeast (Fig. 18). In recent years, new and, in many respects, unexpected information has been obtained through heroic experiments in which the eye of the storm and its vicinity have been sounded by airplanes (see, for example, Simada[123]). Photography from rockets and satellites has also been fruitful. For example, Fett[124] reports the observation of a narrow cloudless zone of descending motion along the boundary of the cloud system of a hurricane. This zone has an unusually low humidity: The dew point deficit Δ equals 15 to 20°. In front of it there is an external zone of intense convection with thick cumulus clouds. A strong jet stream above the settling zone (at a level of about 200 mb) envelops the hurricane anticyclonically from the north and then splits into two branches; the eastern branch curves cyclonically and becomes a trailing whirlwind.

Synoptic methods do not provide a full explanation or an accurate prediction of the motion and evolution of hurricanes. For the time being, the same is true of hydrodynamic methods as well. In par-

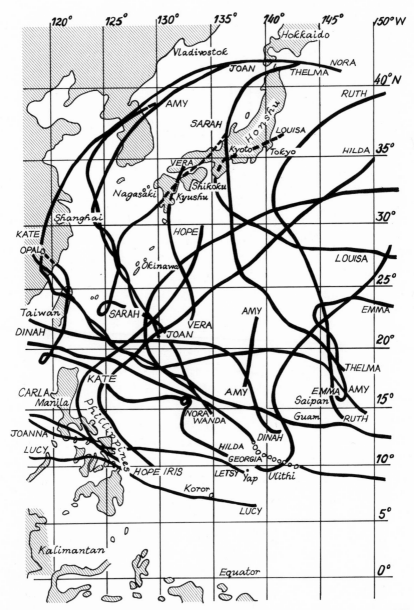

Fig. 18   Paths of typhoons in the western part of the Pacific Ocean for the 1962 season (schematic).

ticular, the sharp turns observed in the motion of hurricanes have still not been explained physically. According to Hill,[125] the U.S. Weather Bureau's daily forecasts of the positions of hurricane centers have an average error of more than 200 km, while only forecasts with errors no greater than 80 km can be considered satisfactory. (The U.S. Weather Bureau's forecasts are so far considered the best.) The simplified hydrodynamic equations that are now used are obviously inadequate in accounting for many factors essential to the evolution of hurricanes: friction, release of the heat of condensation, and possibly the vertical acceleration.

However, some published physicomathematical models of typhoons have given encouraging results. So, for example, Estoque[126] describes calculations according to two models; one is a quasi-gradient model, and the other is based on the primitive equations. The vertical and horizontal turbulent exchange and the release of the latent heat of condensation in upward movements are taken into account, and the nonstationary axially symmetrical problem is solved with time intervals of 90 and 15(!) sec. The results of the calculations reproduce the descent of air in the eye of the storm and its rise at the "wall of the eye," the convergence at the lower levels and divergence at the higher levels.

Morikawa[127,128] constructed a hurricane model in which, by taking its small (mesometeorological) scale into account, the hurricane is treated as a "point vortex" interacting with the large-scale flow. In other words, he proposed to describe a hurricane by using the solution of the quasi-geostrophic barotropic equation (7.6') in the form

$$z = \frac{l}{g} (\psi_0 + \psi_1),$$

where

$$\psi_0 = \frac{\gamma}{2\pi} K_0 \left[\frac{|r - r_0(t)|}{L_0} \right]$$

$[(l/g)\psi_0$ is the solution of equation $(7.6')$ corresponding to a point vortex at the point $r_0(t) = \{x_0(t), y_0(t)\}]$, and where the trajectory of the vortex is determined by the relations

$$\frac{dx_0}{dt} = -\frac{\partial \psi_1}{\partial y}\bigg|_{r_0} \qquad \text{and} \qquad \frac{dy_0}{dt} = \frac{\partial \psi_1}{\partial x}\bigg|_{r_0}$$

This solution was used with some success in describing the motion of hurricane Betsy for 14–17 August 1956; in addition, a model was calculated with a quasi-uniform basic flow stream function

$$\psi_1 = UL_0(e^{-y/L_0} - 1) \approx -Uy.$$

In the papers cited, Morikawa used an approximate representation of the continuous vorticity field (or better, the potential vorticity); his model consisted of a finite number of point vortices whose motion is described by *ordinary* differential equations. This is analogous to replacing a continuous distribution of mass by a discrete set of point masses. This in turn is a way of approximately describing continuous fields that may be an alternative to the use of an expansion in orthogonal functions (section 10) or the use of discrete spatial grids (section 11). In hydrodynamics this method has been used by Onsager,[129] Fermi, Pasta, and Ulam,[130] Pasta and Ulam,[131] and Ulam.[132] For the problems of meteorology, it was advocated by Charney.[133]

To summarize this section, it can be admitted that the hydrodynamic theory to some degree already copes with the short-range forecast not only of the pressure field but also of such weather components as the temperature and wind. It is making important steps toward the prediction of cloudiness and precipitation. The weather services claim that they are already using the hydrodynamic theory of weather forecasting in practice. These claims can be accepted only when objective hydrodynamic forecasts are compiled, not *along with* subjective synoptic forecasts, but *in place* of them.

The Physical Nature of Long-Term Weather Changes

13. The Global Nature of Long-Term Weather Processes

According to the definition given in section 4, time intervals $t - t_0$ are called long if they are large compared to the typical time for the generation of the kinetic energy associated with the synoptic processes $\tau = [(1/E)(\partial E/\partial t)]^{-1}$ [or the energy dissipation time $\tau \sim (1/\varepsilon)(E/M) \sim U^2/\varepsilon$], which is on the order of one week. Over long periods, $t - t_0 > \tau$, all the regions of the atmosphere have time to interact with each other. (For example, large-scale tropospheric perturbations are known to propagate with speeds of up to 35–45 degrees of longitude and 10 degrees of latitude per day; thus they can circle the earth within one to two weeks. Note also that the middle layers of the atmosphere in the middle-latitude west-east flow circle the earth within a period on the order of one month.)

Consequently, for long-term changes the atmosphere works like a single system, all of whose parts strongly interact with each other. No part of the atmosphere can be considered in isolation over long periods of time since, as a result of the interaction, the evolution of one part is determined not only by its own state but also by the states of all the other parts of the atmosphere. (Hence follows, in particular, the fallacy in introducing one of the basic concepts of the so-called synoptic method of long-range weather forecasts—the "natural synoptic region," which comprises Europe and the eastern Atlantic. The state of the atmosphere in this region, it is claimed, determines the long-term weather changes there, independently of the state of the atmosphere in its other parts. In addition, other basic concepts of this method—rhythms and analogs—obviously do not conform to reality.)

In other words, long-term atmospheric processes are inevitably global; for them, only the planet as a whole can serve as a "natural synoptic region." Thus, these processes ought to be viewed as oscillations of the general circulation of the atmosphere, and the dynamic equations of the general atmospheric circulation on a spherical earth should be used for their description (the equations were formulated in 1936 by Kochin[1]).

In climate theory studies are made either of the stationary solutions of these equations (some outstanding and important examples of this are Blinova's[2] construction of the centers of action of the atmosphere, her calculation[3] of the average annual zonal temperature field, and Mashkovich's[4] calculation of the average annual zonal circulation), or of the solutions with a one-year period (Blinova,[5] Blinova and Marchuk[6]), or, finally, of nonstationary solutions with idealized initial data corresponding to small perturbations of the rest state (the so-called numerical experiments on the general circulation of the atmosphere, which will be treated in detail below).

For the purposes of long-range weather prediction, the nonstationary solutions of the very same equations are required but with concrete initial data that fix the initial state of the entire physical system under investigation. This approach to the description of long-term atmospheric processes has been successively developed in a 25-year series of papers by Blinova, beginning with her 1943 paper[2]; her review[7] published in 1957 and her recent summary works[8-12] should be mentioned here.

Note that over long periods of time all parts of the atmosphere succeed in interacting not only with each other but also with the active layer of the underlying surface. Here the most important role must belong to the ocean (since, as a result of the great heat capacity of water compared to soil, the heat content of the active layer of the ocean can be much higher than that of the active layer of soil). According to the estimate given in the report of a 1967 GARP conference,[13] the reaction of the upper layer of the ocean on atmospheric processes begins to be substantial after one to two weeks.

Thus, in long-term processes the atmosphere as a whole does not act as a closed system but only as a component of the single system composed of the atmosphere and the active layer of the underlying surface (this will henceforth be called the A–AL system for short).

In this system the atmosphere is a rapidly changing component of low inertia; but the ocean, on the other hand, has a large thermal inertia (note, for example, that heating the entire mass of the atmosphere an average of 6°C requires only a 0.1°C cooling of a 100-m layer of water). One may therefore think that the initial heat distribution in the active layer of the "World Ocean" can determine the evolution of the atmospheric processes for a fairly long period of time. In other words, the most important initial condition for long-range weather prediction apparently must be the temperature field in the active layer of the World Ocean (see my paper[14]); it is even possible that, in order to lengthen the period of validity of short-range forecasts, it is sufficient to describe the heat exchange between the atmosphere and the ocean under the simplifying assumption that the temperature of the ocean is constant.

14. Global Observation

14.1. In order to predict the evolution of a complex system whose parts all strongly interact with each other, one must obviously be able first of all to fix the initial state of the system as a whole. But regular observations of the atmosphere are now organized just over an area of land that amounts to only about 15% of the earth's surface; and regular observations of the atmosphere above the ocean and of the state of the active layer of the ocean are practically unavailable (except for observations from a few weather ships). Thus we can now fix the initial state of only a small fraction of the A–AL system. This is totally inadequate for the long-range prediction of weather.

One necessary step toward the latter goal is the organization of the global observation of the atmosphere and the ocean. The reason for its necessity is explained by Charney,[15] for example. Because the practical solution of this problem requires very large expenditures, the International Council of Scientific Unions and the World Meteorological Organization decided to first carry out experiments in global observation (apparently they will be ready for 1973–1976;

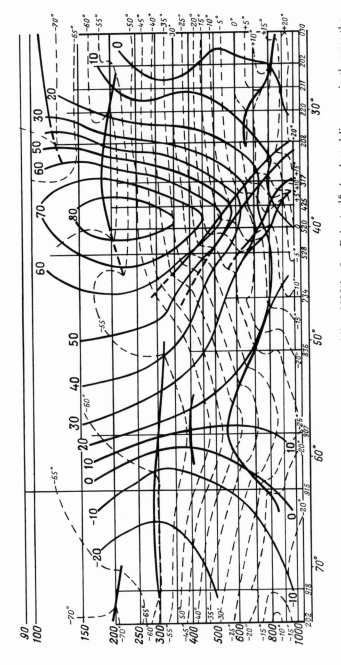

Fig. 19 A vertical section of the atmosphere along the meridian 90° W, after Palmén,[16] the dotted lines are isotherms; the thin solid lines are the isolines of the zonal wind speed (in m/sec); the thick solid lines are the surfaces of a division (fronts and the tropopause).

see the previously cited GARP report[13]); the results will make it possible to plan a future system of regular global observation of the ocean and the atmosphere in the most effective and economical form.

14.2. One may be confident that converting to a more perfect observational system will lead to the discovery of new regularities in the general circulation of the atmosphere. Thus, thanks to the creation of a network of regular aerological soundings of the troposphere and the lower stratosphere, one of the greatest links of the general atmospheric circulation was discovered: the *jet streams* girding the entire globe. Located at the boundary between the troposphere and the stratosphere, the jet streams have a width of only 300–700 km, and their wind velocities reach 80–100 m/sec and even higher. Until the advent of artificial earth satellites, postwar observational meteorology saw no greater discovery.

As an example, Fig. 19 gives a now-classic vertical section of the middle-latitude atmosphere along the 90° W meridian (from Palmén's article[16]); in it one can clearly see the jet stream with a maximum velocity of about 85 m/sec at a latitude of 39° and a level of 200mb. The jet stream was obtained (in a smoothed, average annual form) by theoretical means in Mashkovich's calculation.[4] For a review of the present data on the jet streams, see, e.g., Reiter's monograph.[17] *

14.3. In organizing the global observation of the atmosphere and the ocean, the territories that are as yet least illuminated by observations naturally attract immediate attention. Foremost among these is the *tropical zone*, which apparently plays an important role in supplying the atmosphere as a whole with heat and moisture. Observations of the tropics have so far been quite inadequate. However, Sadler[18] has demonstrated that now, at the cost of great efforts in gathering all available synoptic, aerological, and aircraft observations and satellite data, it is already possible to carry out an operational analysis of the synoptic processes in the tropics twice daily. Such an analysis shows the error in dogmatically assuming the

---

*See also E. N. Lorenz, *The Nature and Theory of the General Circulation of the Atmosphere*, WMO, 1967.

invariability of the weather in the tropics. It follows then that weather forecasts are also necessary in the tropics; as an example, see one of Sadler's maps (Fig. 20).

In tropical meteorology today there are more riddles than solved problems (see Riehl's book,[19] Charney's article,[15] and the 1967 GARP report[13]).* The five following examples of such riddles and problems are possibly among the most important.

1. The trade winds are apparently maintained by the heat given off in the condensation of moisture in the thick convective clouds in the narrow (sometimes only about 100 km in width) intertropical convergence zone (ITCZ). Why is the ITCZ located, not at the equator, but usually at latitudes from 3° to 20°? A preliminary explanation suggested by Charney[20-22] takes into account both the influence of the frictional convergence (which disappears at the equator) at the boundary layer of the trade winds and the influence of the temperature maxima of the ocean surface (which are not symmetrical with respect to the equator because of the rise of cold water—caused by the easterlies—at the equator).

2. What are the typical times for interaction between the atmospheric processes in the Northern and Southern hemispheres? For periods that are short compared to the typical interaction time, it is clearly possible to view the Northern and Southern hemispheres as independent; this would make prediction of the world's weather substantially easier. Litvinenko[23,24] used real data for 1958–1961 to calculate the meridional fluxes of the mass of air across the equator in the layer from 850–100 mb. He established that the air interchange across the equator is intense; it varies irregularly during the year and has a complex spatial structure. According to his data, about one-fourth of the entire mass of the atmosphere M crosses the equator in one month. The resultant transfer is of course much smaller: It is only a fraction of a percent of M per month.

3. In the equatorial stratosphere there is a zonal flow (similar to the equatorial flows in the atmospheres of Jupiter and the sun). This flow occurs within the limits of $\pm 10°$ latitude and at altitudes of up to 30 km. Its vertical profile is a wave with an alternation of westerlies and easterlies in layers with thicknesses on the order of 5 km; the wave travels downward with an average period of about 26 months (see Fig. 21, adopted from Reed's article[25]). Manifestations of a 26-month rhythm are present in other regions of the atmosphere in the various meteorological fields as well. What is the reason for these 26-month oscillations?

4. The main synoptic perturbations in the trade zone are the easterly waves (with lengths on the order of 1000–2000 km) with cold centers. Hurricanes have warm

*See also M. A. Alaka, *ESSA Tech. Rep. WB-6* (May 1968), 18 pp.; G. W. Cry and W. H. Haggard, *Monthly Weather Rev.* **90**, 341–349 (1962); R. R. Dickson, ibid. **97**, 830–834 (1969); N. L. Frank, ibid. **97**, 130–140 (1969); N. L. Frank, ibid. **98**, 307–314 (1969); W. M. Gray, ibid. **96**, 669–700 (1968); R. H. Simpson, N. Frank, D. Shideler, and H. M. Johnson, ibid. **97**, 240–255 (1969).

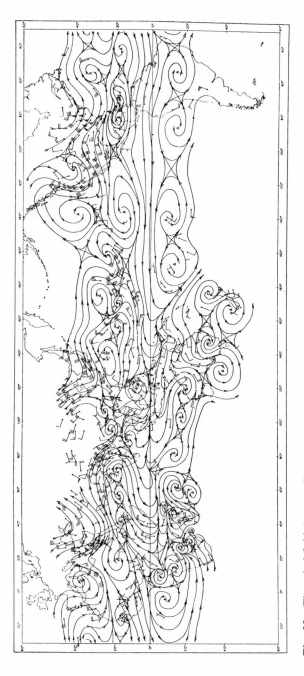

Fig. 20 The wind field (current lines) at a level of 700 mb in the tropic zone on 10 December 1963 (constructed from 300 bits of initial information). After Sadler.[18]

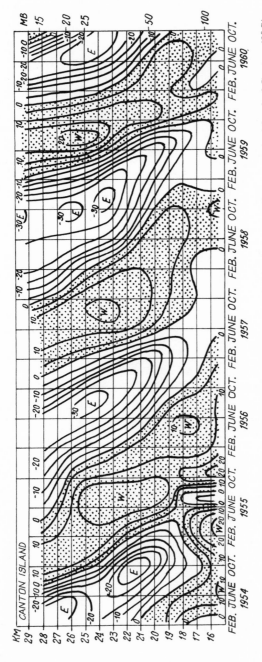

Fig. 21 Altitude-time isopleths of the vertical profile of the zonal wind in the stratosphere over the island of Canton (3° S); speed in m/sec. After Reed.[25]

centers, but they most often originate in easterly waves. How are easterly waves with cold centers formed, and why do they turn into hurricanes with warm centers?

5. The Coriolis parameter l vanishes at the equator, and the Ekman boundary layer [with a thickness on the order of $(K/l)^{1/2}$] cannot exist. What is the structure of the boundary layer of the atmosphere in the equatorial zone? It is possible that inertial forces play an important role in it and that its structure varies in the horizontal direction as in the boundary layer for the flow around a flat plate or as in the boundary layer in hurricanes.

14.4. Artificial earth satellites are a powerful new means of global observation. According to the discussion of the theoretical possibilities for such use of satellites by Malkevich, Monin, and Rozenberg,[26] the only available carrier of meteorological information is the field of electromagnetic radiation (of various wavelengths) that is emitted, reflected, or transmitted by the atmosphere and the earth's surface. Therefore our first task is discovering the regular relationships between, on the one hand, the spatial, angular, and temporal structure of this field and, on the other hand, the structure of the various meteorological fields and the underlying surface. Evidently this problem is most easily solved by satellite observation (telephotography) of the cloud field in visible light.

This turns out to be a very fortunate circumstance since cloud cover is one of the most important weather elements. In the first place, it directly affects people and influences their activities: It screens the sun's radiation, determines the lighting conditions, influences the daily variations of the temperature, turbulence, and wind, carries precipitation, determines the flying conditions, and so on. Second, the cloud field serves as an effective regulator of the fundamental processes that form long-term weather changes: the influx of solar heat to the active layer of the ocean and the transfer of heat from the ocean to the atmosphere (this will be dealt with below). Third, the cloud cover can serve as an indicator of several other meteorological fields: the wind at the cloud level, peculiarities of the temperature stratification of the troposphere (inversion layers and regions of convective instability), the humidity field, and the ongoing synoptic processes (particularly atmospheric fronts).

The importance of the cloud cover as a weather element and the relative simplicity of observing it by satellite are responsible for the success of the first weather satellites: the Tiros series (the first of these was launched on 1 April 1960, the day meteorologists jokingly call "International Weatherman's Day"). Their primary task was the telephotography of clouds and the measurement of infrared radiation in several spectral intervals, including the "window of transparency" from $8-13\,\mu$. Summaries of the results obtained by Tiros can be found, e.g., in the articles by Kletter[27] and Lariviere[28] (see also the review books by Kondrat'yev[29, 30]).

The large-scale (thousands of kilometers), mesoscale (from one to hundreds of kilometers), small-scale (from tens of meters to kilometers), and microscale (from millimeters to tens of meters) features of the cloud cover in satellite photographs (as well as different-scale features of other fields registered by satellites) should be analyzed in different ways, according to the article by Malkevich, Monin, and Rozenberg.[26] For the *large-scale* features this analysis makes it possible, in the first place, to study the statistics of the cloud field. For example, an analysis of 1447 satellite photographs enabled Arking[31] to construct the average cloud distribution for the zone lying within $\pm 60°$ latitude during the period from July to September 1961; here it will be important to discover long-standing anomalies of the large-scale cloud field. Second, of course, the analysis of individual large-scale cloud systems is of great interest. For example, in Jones's article [32] the four-day evolution of an extratropical cyclone is traced, using *Tiros 1* photographs, from its greatest development to its death. In many cases satellite photographs make possible substantial corrections in the analysis of synoptic charts (see Fig. 22 for an example from Hubert's article[33]). Another example is Winston:[34] Typhoon Debby was located by means of *Tiros 3* photographs for 10 September 1961, after it had been found by weather forecasters several days earlier and had then been lost. Satellite photographs have proved to be an effective means of tracking hurricanes: *Tiros 3* alone made it possible to discover and track 18 typhoons.

Perhaps most sensational has been the discovery of several *meso-scale* peculiarities of cloud structure; these are small in comparison to the distances between weather stations and therefore are not caught in the fabric of the meteorological net, but they are at the same time too large to be observed from a single station. They are, first, the *spiral structure* of cloud systems of large-scale atmospheric vortices and cyclones, then the long, narrow *cloud streets*, and finally, the *cell structure* of cumulus clouds. For an example of the latter see the papers by Krueger and Fritz[35] and by Priestley[36] on the cell structure of cumulus convection; it has been established that air

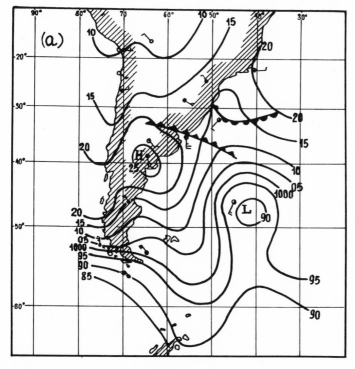

Fig. 22a A map of the surface pressure field for 12^h 28 April 1960, the analysis given by the Argentine weather service.

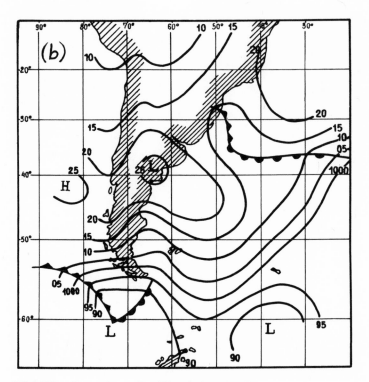

Fig. 22b Correction, using *Tiros 1* photographs.

settles in the centers of the cells and rises on their peripheries; the horizontal diameters of the cells are on the order of 50–80 km, which is thirty times larger than their heights—a ratio ten times larger than that for convective cells in laboratory experiments. Another curious example is the observation from *Tiros 3* of a cloud of locusts in eastern Africa.[37]

It is advisable to describe the random *mesostructure* and also the *small-scale structure* of satellite observations (created mostly by the atmospheric turbulence) statistically, let us say, by using the spatial correlation functions.[26] Finally, the *microstructure* can be produced by an ocean wave, for example, and can be studied using the character of the sun's reflection from the ocean surface (the "sun glint") and polarization effects.

The telephotography of clouds and the measurement by satellite of radiation coming off clouds have been substantially improved in recent years by means of (1) the improvement of camera equipment and infrared detectors; (2) the orientation of satellites relative to the earth's surface, reducing the perspective distortions in the photographs; (3) the launching of satellites in polar or near-polar orbits, allowing observation of the whole planet from pole to pole (e.g., satellites of the Nimbus, Cosmos, and ESSA series); (4) the launching of synchronous satellites in an equatorial orbit at an altitude of 36,000 km, where their period of revolution is exactly equal to one day [in this way, the satellites are motionless with respect to the surface of the earth (for example, *ATS 1*, above the middle of the Pacific Ocean), allowing continuous observation of the entire hemisphere and the exact measurement of cloud movements].

An example of a recent cloud photograph obtained from *ESSA 5* (and kindly provided by S. Fritz from ESSA) is given in Fig. 23. In this picture, which completely covers the zone from 30° S to 40° N, all the tropical cloud systems are perfectly visible. Special attention is drawn to the cloud cover of the ITCZ (which occupies the zone from 0–10° N over Africa, a narrower zone near 10° N over the Atlantic, and a broken zone near 10° N over the Pacific; it doesn't appear over the Indian Ocean), the spiral cyclonic structures over southern Japan, to the east of it, and west of California, the orographic cloud cover over the Cordilleras in South America, and the very long cloud belts over Indochina. It is clear that the regular reception of such photographs creates an entirely new basis for tropical meteorology, for example.

Satellites now make it possible[13] to fix the cloud cover with a tolerance of 1 × 1 km during the day and 10 × 10 km during the night and from its motions to make a rough determination of the wind velocity at cloud level. Satellites can also measure outwardly traveling radiation in the spectral band from 4–30 μ with a tolerance of 5 × 5 km and the temperature of the surface of the earth or the upper boundary of the clouds (by infrared radiometry) with an accuracy of $\pm 1°$ and a tolerance of 10 × 10 km. However, as was indicated above, in weather forecasts it is first of all necessary to have measurements of the three-dimensional pressure and

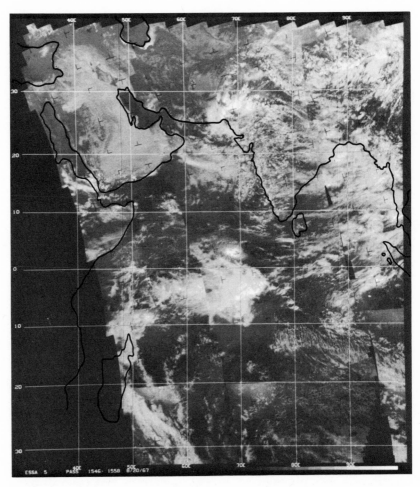

Fig. 23a A photograph of the cloudiness in the tropic zone between 40° E and 100° E (obtained from the satellite *ESSA 5*).

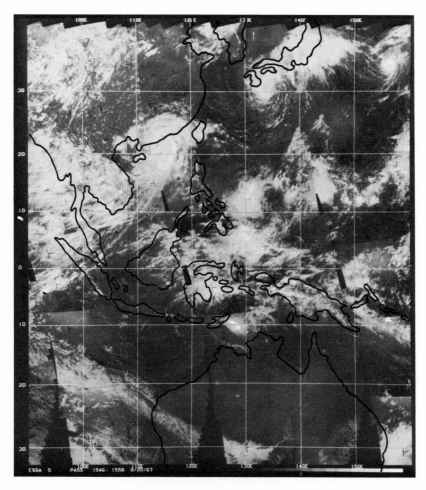

Fig. 23b A photograph of the cloudiness in the tropic zone between 100° E and 160° E (obtained from *ESSA 5*).

Fig. 23c   A photograph of the cloudiness in the tropic zone between 160° E and 140° W (obtained from *ESSA 5*).

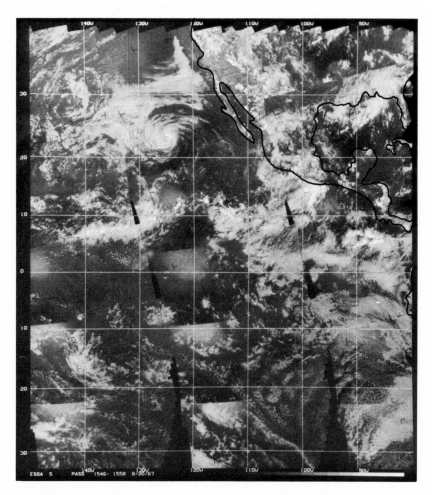

Fig. 23d A photograph of the cloudiness in the tropic zone between 140° W and 80° W (obtained from *ESSA 5*).

Fig. 23e   A photograph of the cloudiness in the tropic zone between 80°W and 20°E (obtained from *ESSA 5*).

Fig. 23f   A photograph of the cloudiness in the tropic zone between 20° W and 40° E (obtained from *ESSA 5*).

wind fields. Thus, the question, How should satellite data be used for the purposes of weather forecasting? arises. There are now several articles and even books devoted to attempts to answer this question (for example, see the book by Kondrat'yev, Borisenkov, and Morozkin[38]).

Of course, it is more constructive to search, not for problems that can be solved by a given method, but for methods by which a given problem can be solved. But sometimes technique outstrips theory, and essentially new technical means such as radioisotopes, lasers, and satellites appear before their primary destination is determined. And although today's satellite data are undoubtedly useful for diagnosing the world's weather, a highly progressive step would be the development of satellite measurements of the meteorological fields that are directly necessary for hydrodynamic weather prediction.

In this connection the following possibilities are now under discussion (e.g., at the 1967 GARP conference[13]):*

1. Determination of the vertical profiles of the temperature (with an accuracy of $\pm 2°$) and the relative humidity (with an accuracy of $\pm 10\%$) *above the clouds.* Data could be used from a multichannel infrared spectrometer with a spectral tolerance of $5 \, \mathrm{cm}^{-1}$ in the region of the $15 \, \mu \, CO_2$ absorption band (so far, infrared spectrometry experiments with a $30 \, \mathrm{cm}^{-1}$ tolerance have been carried out[39, 40] on *Tiros 7* and *Nimbus 2*), and the determination could be made by solving the reverse problem, e.g., after Wark and Fleming,[41] with the aid of empirical orthogonal functions. The possibility of taking measurements *through clouds* by microwave spectrometry is also under discussion.

2. Determination of the vertical density profile (with an accuracy perhaps better than $\pm 1\%$ and with an averaging over a volume of $1 \times 1 \times 400 \, \mathrm{km}$). The method of microwave radioscopy of the atmosphere could be used with rays traveling along different chords from a satellite transmitter to five or six satellite receivers located in the same orbit. (A similar method was used for determining the density of the Martian atmosphere in a *Mariner 4* experiment.)

3. Measurements of the wind field above clouds. Krause has suggested a "cross-beam" method, i.e., the simultaneous observation of a small volume of the atmosphere by two separated radiometers and separating out the correlated components in both signals.

However by no means can or should all measurements be transmitted by satellite; on the contrary, for most measurements it is necessary, now as before, to place transmitters *in situ*, i.e., within the atmosphere or ocean. For the organization of global observations, it

*See also *GARP Publ. Series*, No. 1 (Oct. 1968), No. 2 (Jan. 1969), No. 3 (Oct. 1969), No. 4 (Jan. 1970), and *COSPAR Working Group 6* (Feb. 1970).

will therefore be necessary to reinforce the net of weather stations on the continents and to create a network of stationary (anchored) buoy stations on the oceans. During experiments in global observation as a temporary measure it will be possible to use mobile measuring platforms—super-pressure probe balloons in the atmosphere and floating buoy stations in the ocean; information concerning the platforms can be gathered primarily by satellite; moreover, it will be necessary for the satellites to determine the location of the interrogated mobile platform simultaneously with the interrogation with sufficient accuracy.

An especially important task is the creation of a network of stationary buoy stations on the oceans; thus, for example, the U.S. Weather Bureau intends[43] to begin with a network of 300 buoy stations interrogated by three synchronous satellites. This will make detailed instrumental observations possible on a large portion of our planet and finally over its entire surface; in particular, it will permit the accumulation of the continuous and lengthy series of measurements needed for studying the processes responsible for the variability of the oceanological fields and the atmosphere over the oceans. An experimental network of seven buoy stations, measuring the temperature and current-velocity profiles, was set up for two months during the fortieth voyage of the scientific research ship *Vityaz* (in 1967 in the Indian Ocean); it showed a large spatial and temporal variability even in what would seem to be a relatively calm region of the ocean.

A system of global observation of the atmosphere and the ocean promises to yield an immense amount of information—according to Singer,[44] on the order of 10^{10} bits per day; by way of comparison, a modern synoptic chart of the Northern Hemisphere contains 10^5 bits, and a satellite photograph of the cloud cover contains 10^7 bits of information. Processing and storing this information will require computers and information machines with high-speed operation and large-capacity memory devices.

15. Nonadiabatic Effects

According to the definition of long periods given in sections 4 and 13, the most important thing in the analysis and prediction of the long-term evolution of the world's weather must be the description of non-adiabatic factors—sources and dissipations of energy. The fundamental equations (4.1) of the theory of short-term atmospheric processes, expressing the laws for the *conservation* of the entropy s and the potential vorticity Ω must now be replaced by equations for the *evolution* of the entropy and potential vorticity:

$$T\frac{ds}{dt} = \varepsilon, \qquad \rho\frac{d\Omega}{dt} = \left(\Omega_a \cdot \nabla \frac{\varepsilon}{T}\right) + (\text{curl}\,\mathbf{f}\cdot\nabla s), \qquad (15.1)$$

where ε is the rate of energy increase per unit of mass (having the dimension of velocity squared divided by time), and $\mathbf{f}$ is the viscous force per unit of mass. The first of these equations is the usual equation for the heat increase. The second, the equation for the evolution of the potential vorticity, was obtained by Obukhov,[45] who also pointed out that its right side can be written in the form

$$\text{div}\left(\frac{\varepsilon}{T}\Omega_a + s\,\text{curl}\,\mathbf{f}\right),$$

so that the solenoidal character of the field $(\varepsilon/T)\Omega_a + s\,\text{curl}\,\mathbf{f}$ serves as the general condition for potential vorticity conservation.

Gavrilin[46] pointed out that with the quasi-static approximation and when the horizontal component of the curl of the earth's rotation in the Coriolis acceleration equations is, as is customary, neglected, the second equation of (15.1) remains valid if the expressions of the form $-(1/g\rho)\,(\text{curl}\,\mathbf{A}\cdot\nabla B)$ are rewritten: e.g., as

$$(\text{curl}\,\mathbf{A})_z\frac{\partial B}{\partial p} + \frac{1}{a\sin\theta}\frac{\partial A_\theta}{\partial p}\frac{\partial B}{\partial \lambda} - \frac{1}{a}\frac{\partial A_\lambda}{\partial p}\frac{\partial B}{\partial \theta}$$

in the coordinates θ, λ, and p, where $\theta = \frac{1}{2}\pi - \varphi$ is the complement of the latitude φ, increasing toward the south, and λ is the longitude, increasing toward the east. With the quasi-solenoidal approximation, in expressions of this form it is apparently sufficient to keep only the first term, and the equation for the evolution of the potential vorticity can be reduced approximately to the form[46]

$$\frac{\partial F}{\partial t} + [\psi, F] = -R\frac{\partial}{\partial p}\frac{p}{\alpha_0^2 c_0^2}\frac{\varepsilon}{c_p} + \frac{\Omega_f}{\nabla^2\psi + l'}, \qquad (15.2)$$

where F is the approximate adiabatic invariant (8.2), and $\Omega_f \approx D\{\nabla^2\psi\}$ is the vertical projection of the curl of the frictional forces; here D is the usual diffusion operator, which in the coordinates θ, λ, and p is often given in the form $D = K_h\nabla^2 + (\partial/\partial p)\,(p^2/\tau_v)\,(\partial/\partial p)$, where K_h is the coefficient of horizontal mixing, and τ_v is the "vertical mixing time." In the atmosphere, the nonadiabatic heating ε is determined primarily by the temperature and humidity fields (this statement will be made somewhat more precise below). Consequently, equation (15.2) along with the balance equation and the equations for the humidity field would theoretically make it possible to determine the fields z and ψ and the humidity field in the atmosphere. The magnitude of w^* could then be determined from equation (12.1) with the term $-\mathfrak{U}\,(\partial\Omega_f/\partial p) - (R/p)\,(\nabla^2\varepsilon/c_p)$ added to its right side if the boundary conditions at the limits of the atmosphere did not contain any other unknown functions.

Let us deal briefly with the boundary conditions. At the upper boundary of the atmosphere $(p \to 0)$ it is sufficient to require some regularity in the fields z and ψ, and it is necessary to fix the incoming solar radiation flux (which is determined from astronomical data). On the lower boundary of the atmosphere $(p \approx p_0)$, several requirements are necessary.

First, on a dry-land surface the horizontal component of the velocity must go to zero, and on the surface of the ocean there must be continuity in the vertical flux of the horizontal component of momentum (more precisely, the difference between such fluxes in the air and in the water must equal the rate of increase of the total momentum of the wind waves generated in the thin upper layer of the sea).

Second, the vertical velocity w must go to zero; and in calculating its value w_0, which arises from variations of the relief and the frictional convergence in the boundary layer of the atmosphere, condition (12.2) must be satisfied. By expressing w^* in terms of the fields z, ψ, and ε with the equation

$$w^* = \frac{gp^2}{{\alpha_0}^2 {c_0}^2}\left(\frac{d_h}{dt}\frac{\partial z}{\partial p} + \frac{R}{gp}\frac{\varepsilon}{c_p}\right),$$

which is equivalent to the first equation of (15.1), we can write this condition in the form

$$\frac{d_h}{dt}\left(p\frac{\partial z}{\partial p} + {\alpha_0}^2 z\right) = {\alpha_0}^2 {w_0}^2 - \frac{R}{g}\frac{\varepsilon}{c_p} \quad \text{at} \quad p = p_0. \qquad (15.3)$$

Third, the algebraic sum of all vertical heat fluxes must go to zero at the lower boundary of the atmosphere. Five kinds of flux enter into this sum : (1) the heat flux in the active layer of the ocean or the soil (produced in water by turbulence as well as the transfer of short-wave radiation, and in soil by its heat conductivity), (2) the turbulent heat flux in air, (3) the turbulent flux of latent heat (the product of the heat of vaporization $\mathscr{L}$ and the rate of evaporation, i.e., the turbulent flow of water vapor), (4) the short-wave radiation flux in the air, and (5) the long-wave radiation flux in the air. Methods of presenting these fluxes are given in detail in Kibel's book[47] and Blinova's papers,[7-12] for instance.

The momentum and heat fluxes in the active layer of the soil or the sea entering into the boundary conditions at the lower boundary of the atmosphere (as well as the temperature of the underlying surface, which determines its long-wave radiation and the evaporation conditions) are new unknowns, added to the atmospheric fields z and ψ and the humidity field. For their calculation, equations describing the processes in the active layer of the underlying surface must be called into play: in soil, the distribution of heat (and moisture, perhaps), and in the ocean, the movement and distribution of heat and salinity.

16. Heat Sources in the Atmosphere

We have already considered the heat sources in the atmosphere ε in connection with equation (12.6) [which is equivalent to the first equation of (15.1)], and we noted that ε consists of three terms: (1) the heat increase due to the turbulent heat conductivity of air, usually written in the form $-1/\rho \operatorname{div} \mathbf{Q}_T \approx c_p D\{T\}$, but, according to my paper,[48] here it is more correct to take, instead of the turbulent heat flow $\mathbf{Q}_T$, the turbulent entropy flux multiplied by T, i.e., the quantity $\mathbf{Q}_T + [(s_v - s_a)/c_p](c_p/\mathscr{L})\mathbf{Q}$, where s_a and $\hat{s}_v$ are the entropy densities of dry air and water vapor, and $\mathbf{Q}$ is the turbulent flux of latent heat; (2) the heating accompanying phase changes of moisture $\mathscr{L}m$, where m is the specific rate of condensation (plus sublimation) of water vapor, nonzero only in clouds, where it is given, for example, by Shvets's equation (12.8) [since the latter contains w^*, the unknown function w^* also appears in the right side of equation (12.1)]; (3) the radiation heat flux divergence ε_r, which in turn consists of the two terms ε_{rS} and ε_{rL}, determined respectively by the short-wave radiation of the sun and the long-wave radiation of the underlying surface and the atmosphere.

Ways of writing the radiative heat changes in the equations of the hydrodynamic theory of weather forecasts have been studied in detail in Kibel's book[47] and in Blinova's papers.[7-12] We shall deal with these here, trying first to go from their general definition to constructive formulas for their computation and second to show an obvious way to taking the cloud cover into account. To begin with, let us represent the radiative heat flux divergence in the form $\varepsilon_r = -(1/\rho) \operatorname{div} \mathbf{F}$, where $\mathbf{F}$ is the radiative energy flux. Since the large-scale horizontal changes in the field $\mathbf{F}$ are much smaller than the vertical changes, it is quite permissible to put $\varepsilon_r \approx - (1/\rho)\, (\partial F_z/\partial z)$ or

$$\varepsilon_r = g\frac{\partial}{\partial p}(F^\uparrow - F^\downarrow),\tag{16.1}$$

where $F^\uparrow$ and $F^\downarrow$ are the upward and downward *vertical* radiative energy fluxes. In calculating these fluxes in the *earth's atmosphere*, several simplifying assumptions are made: (1) The lower 50-km layer of the atmosphere is assumed to be in a state of *local thermodynamic equilibrium* (i.e., every infinitesimal volume of air radiates and absorbs radiation as a perfectly black cavity of the same temperature and in thermodynamic equilibrium). Therefore, in particular, Kirchhoff's law holds: The ratio of the spectral coefficients of radiation and absorption does not depend on the nature of the optically active substances and is a universal function of the wavelength and temperature (Planck's function). (2) The *polarization* of the radiation is not taken into account (this may be unsatisfactory for short-wave radiation). (3) In the general case, *refraction* is not taken into account (it can only be important in special cases). (4) Sometimes *dispersion* is not taken into account (it does not contribute directly to div $\mathbf{F}$, and in the case of long-wave radiation it is usually small).

We shall first deal with short-wave radiation. The difficulties that arise here will be demonstrated in a scheme for calculating ε_{rs} recommended by the Laboratory for the Mathematical Modeling of the General Circulation of the Atmosphere and Ocean of the Institute of Oceanology, U.S.S.R. Academy of Sciences. For simplicity, let us limit the account to the case of a single-layered cloud cover, the magnitude of which (the fraction of the sky covered by clouds) will be signified by the letter n. The downward short-wave radiation flux *below the clouds* will then be written in the form

$$F_S^\downarrow(p) = \gamma\{(1-n)\,[S - Q_0(p)] + n(1 - \Gamma - \Pi)\,[S - Q_1(p)]\},\tag{16.2}$$

and *above the clouds* it will be given by the same equation, but with $n = 0$. Here $\gamma = \cos \zeta$ (during the day) or 0 (at night), where ζ is the zenith angle of the sun; in calculating the average daily value of ε_{rS} it is necessary to choose for γ the average value of $\cos \zeta$ for the daylight time of the day, multiplied by the relative fraction of daylight time during the day (such values for γ at different latitudes and seasons have been tabulated by Manabe and Möller[49]). Further, S is the value of the solar heat flux at the upper boundary of the atmosphere (the so-called *solar constant*), which is reduced by ozone absorption and Rayleigh scattering (whose contribution to the planetary albedo is taken to be 0.07), after which a solar heat flux of 1.946 cal/cm^2-min is left; Γ is the albedo of the clouds (the ratio of the reflected flux to the impinging flux), and Π is the absorption coefficient in clouds (the ratio of the absorbed flux to the impinging flux); both Γ and Π have been tabulated for clouds of various levels by Manabe and Strickler.[50] Finally, the functions $Q_0(p)$ and $Q_1(p)$ describe the absorption of short-wave radiation by the atmospheric gases in a clear sky and in total cloudiness. The first term within the braces in (16.2) corresponds to the radiation flux passing through the clouds (with some losses).

The main difficulty in calculating the radiant energy fluxes in the atmosphere lies in the complexity of the absorption spectra of the optically active atmospheric gases: water vapor and carbon dioxide. Attempts to avoid taking the absorption spectra into account and to use only a "gray radiation" approximation (for instance, the Schwarzschild-Emden scheme presented in Kibel's book[47]) in calculating the characteristics of the radiative energy field have apparently not succeeded (nevertheless, the gray radiation approximation has been newly proposed in Marchuk's recent paper[51] and in his book[52]—so far, of course, without any concrete calculations). It is therefore necessary to set

$$Q_i(p) = \sum_{j=1}^{3} \sum_{k_j} A_i^j(k_j) S(k_j), \qquad (16.3)$$

where the index $j = 1$ corresponds to the absorption by water vapor, $j = 2$ corresponds to absorption by carbon dioxide, and $j = 3$ corresponds to absorption in the overlapping spectral bands of water vapor and carbon dioxide; k_j is the band number in the jth spectrum; $S(k_j)$ is the amount of radiative energy of the solar flux S in the spectral band k_j; $A_i^j(k_j)$ is the *absorptivity* (the ratio of the absorbed flux to the impinging flux) of the jth gas in the band k_j, with $A_i^3(k)$ set equal to $A_i^1(k) + A_i^2(k) - A_i^1(k)A_i^2(k)$ for $j = 3$.

The absorptivity of the jth gas for monochromatic radiation with wavelength λ depends on the optical thickness of the layer of air through which the radiation passes:

$$\int \rho_j \alpha_\lambda^j \, dz = \frac{1}{g} \int \frac{\rho_j}{\rho} \alpha_\lambda^j \, dp,$$

where α_λ^j is the absorption coefficient per unit mass of the absorbing gas, and ρ_j is the density of the gas. If the dependence of the absorption coefficient on pressure is given by the equation $\alpha_\lambda^j \sim (p/p_0)^{2N_j}$ (p_0 is the surface pressure; usually we set $N_1 = \frac{1}{4}$ and $N_2 = \frac{2}{5}$), then it is clear that the absorptivity will depend on the "effective mass" of the absorbing gas, which is given by the equation

$$m_j(p) = \frac{1}{\rho_j^0 g} \int\limits_p^{p_0} \frac{\rho_j}{\rho} \left(\frac{p}{p_0}\right)^{2N_j} dp, \qquad (16.4)$$

where, for convenience, the standard density of the jth substance ρ_j^0 is placed in the denominator, so that m_j has the dimension of length. In calculations of the radiative fluxes, it is usually supposed that the partial pressure of carbon dioxide is everywhere constant (this assumption clearly saves us from the necessity of including the carbon dioxide concentration field among the meteorological fields to be determined from the dynamic equations); if this partial pressure is taken to be 3×10^{-4}, then the effective mass of the carbon dioxide (which is proportional to the partial pressure) will be equal to

$$m_2(p) = 146[1 - (p/p_0)^{1.8}] \text{ cm}.$$

Note, however, that assuming a constant carbon dioxide content in the air is possibly too crude. The absorption of CO_2 by photosynthesizing plants is carried out at hourly rates that sometimes reach 30–100 mg of CO_2 or more on a horizontal area of 1 dm^2 (Nichiporovich[53]); thus the CO_2 content is sharply reduced in the lowest layer of air (see Fig. 24, adopted from the paper by Chapman, Gleason, and Loomis[54]), and plants experience a shortage of carbon dioxide, so that it is advisable to "fertilize" them artificially with carbon dioxide.[53] The difference in CO_2 content makes carbon dioxide a convenient indicator for tracking air movements and atmospheric mixing. By such a method, Bolin and Keeling[55] established, for example, that the primary natural sources of carbon dioxide in the atmosphere are the tropical regions of the oceans, and the primary industrial sources are the

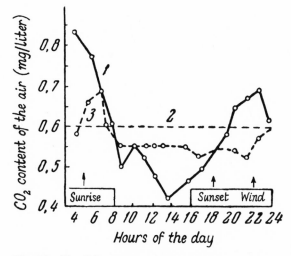

Fig. 24   The CO_2 content of the air at different hours of the day: (1) within a corn field; (2) above clear land; (3) at an altitude of 152 m above a corn field. After Chapman, Gleason, and Loomis.[54]

middle latitudes; moreover, in one year 2×10^{10} tons of CO_2 is carried from the tropics to the polar regions of the Northern Hemisphere (the estimate $K_h = 3 \times 10^{10}$ cm^2/sec is obtained for the coefficient of meridional mixing), and the vegetation north of $45°$ N absorbs about 1.5×10^{10} tons of CO_2 during the summer vegetative season.

To represent the dependence of the absorptivities $A_i^j(k_j)$ on their corresponding effective masses, Shifrin and Avaste[56] constructed the empirical equation

$$A_i^j(k_j) = \frac{x_i^j}{a_j(k_j)x_i^j + b_j(k_j)},\qquad(16.5)$$

where $a_j(k_j)$ and $b_j(k_j)$ are the tabulated spectral coefficients, and x_i^j are simple functions of the effective mass, in this case defined by the formulas

$$\left.\begin{aligned}x_0^j &= (0.76\,p_0)^{N_J}\{\sec \zeta\,[m_j(0) - m_j(p)]\}^{1/2},\\x_1^j &= (0.76\,p_0)^{N_J}\{\sec \zeta\,[m_j(0) - m_j(p_1)] + 1.66\,[m_j(p_1) - m_j(p)]\}^{1/2},\end{aligned}\right\}\qquad(16.6)$$

where p_1 is the pressure at the cloud level (the coefficient 1.66 here takes into account the diffusion of the radiation passing through the clouds).

The upward short-wave radiation flux $F_S^{\uparrow}(p)$ *below the clouds* is composed only of radiation reflected from the earth's surface. It is determined by an equation that differs from (16.2) by a factor equal to the albedo of the underlying surface for short-wave radiation Γ_0 (over the oceans $\Gamma_0 \approx 0.1$, and over land, roughly, $\Gamma_0 \approx 0.2$ in the summer and 0.7 in the winter) and by the replacement of $Q_0(p)$ and $Q_1(p)$ by the functions $Q_2(p)$ and $Q_3(p)$, for which

$$\left. \begin{aligned} x_2{}^j &= (0.76 p_0)^{N_j} \{\sec \zeta m_j(0) + 1.66 m_j(p)\}^{1/2}, \\ x_3{}^j &= (0.76 p_0)^{N_j} \{\sec \zeta [m_j(0) - m_j(p_1)] + 1.66 [m_j(p_1) + m_j(p)]\}^{1/2}. \end{aligned} \right\} \tag{16.7}$$

Above the clouds, to the indicated formula for $F_S{}^\uparrow(p)$ we must first add the factor $1 - n(\Gamma + \Pi)$, which describes the losses of radiation coming from below in its passage through the clouds, and second, the term $\gamma n \Gamma [S - Q_4(p)]$, which describes the flux reflected from the upper surface of the clouds, for which

$$x_4{}^j = (0.76 p_0)^{N_j} \{\sec \zeta [m_j(0) - m_j(p_1)] + 1.66 [m_j(p) - m_j(p_1)]\}^{1/2}. \tag{16.8}$$

The case of a multilayered cloud cover can be described by similar but more cumbersome equations.

Let us now turn to the calculation of the long-wave radiation fluxes. It is appreciably simpler than the calculation of the short-wave radiation flux because, in the first place, it is in fact possible to ignore not only the polarization but also the scattering. As a result, the spectral equation for radiation transfer is easily integrated analytically. After summing with respect to the corresponding wavelengths, its solution is expressed simply in terms of the so-called integral *transmission function* of the diffused radiation

$$D(m_1, m_2) = \frac{1}{\pi} \int \cos \theta \, d\omega \frac{\int E_\lambda(T) \exp [-\tau_\lambda \sec \theta] \, d\lambda}{\int E_\lambda(T) \, d\lambda}, \tag{16.9}$$

where the integration with respect to $d\omega$ is carried out over all directions within the limits of the hemisphere $\theta \le \pi/2$, $E_\lambda(T)$ is Planck's function, and $\tau_\lambda = \int (\rho_1 \alpha_\lambda{}^1 + \rho_2 \alpha_\lambda{}^2) \, dz$ is the *optical thickness* of the air layer through which the radiation passes, expressed in terms of the effective masses m_1 and m_2 of equation (16.4); the transmission function (16.9) was calculated in Nylisk's paper.[57] Second, for long-wave radiation the cloud surface can be considered approximately to be a blackbody, as is shown by a more detailed study presented in Feygel'son's book;[58] however, upper-level clouds should be considered, roughly speaking, to be translucent, so that the corresponding cloud index must be divided by two. Thus, again limiting the discussion to the case of a single-layered cloud cover at a level p_1, we have for the downward long-wave radiation flux *below the clouds*

$$F_L^{\downarrow}(p) = (1 - n)f_{\infty}^{\downarrow}(p) + nf_1^{\downarrow}(p), \tag{16.10}$$

instead of (16.2); and *above the clouds* $F_L^{\downarrow}(p)$ will be determined by the same formula, but with $n = 0$. Here

$$f_{\infty}^{\downarrow}(p) = \int_0^p \sigma T^4(p')\, dD[m_1(p') - m_1(p), m_2(p') - m_2(p)] \tag{16.11}$$

is the downward long-wave radiation flux in a clear sky ($\sigma = 0.826 \times 10^{-10}$ cal cm^{-2} min^{-1} deg^{-4} is the Stefan-Boltzmann constant), and the downward flux from the lower cloud boundary $f_1^{\downarrow}(p)$ is obtained from (16.11) by replacing the lower limit of integration $p' = 0$ by $p' = p_1$ and adding the term $\sigma T^4(p_1)D[m_1(p_1) - m_1(p), m_2(p_1) - m_2(p)]$ to the entire expression. The upward flux of long-wave radiation *below the clouds* will be equal to the flux from the underlying surface

$$f_0^{\uparrow}(p) = \sigma T^4(p_0)D[m_1(p), m_2(p)]$$

$$- \int_p^{p_0} \sigma T^4(p')\, dD[m_1(p) - m_1(p'), m_2(p) - m_2(p')], \tag{16.12}$$

and *above the clouds* it will be expressed by the equation

$$F_L^{\uparrow}(p) = (1 - n)f_0^{\uparrow}(p) + nf_1^{\uparrow}(p), \tag{16.13}$$

where $f_1^{\uparrow}(p)$ is the upward flux from the upper cloud boundary obtained from (16.12) by replacing the first term by $\sigma T^4(p_1)D[m_1(p) - m_1(p_1), m_2(p) - m_2(p_1)]$ and replacing the upper limit of integration p_0 in the second term by p_1 (in a calculation of radiation absorption by water vapor alone, a formula for ε_{rL} similar to the one given here can be found in Feygel'son's paper[59]).

For more precision, in all the calculations of this section it is not hard to separately introduce the levels of the lower and upper cloud boundaries instead of the cloud level p_1.

Until now, in calculating ε_r of the optically active components of the atmospheric air we have taken into account the water vapor (whose concentration forms one of the meteorological fields that are to be predicted) and the carbon dioxide (whose concentration, it is so far considered, can be set equal to a constant). We have in addition roughly taken the ozone into account (in the atmosphere it is concentrated predominantly at altitudes of 20–25 km, where it forms as a result of photochemical processes under the influence of the sun's ultraviolet radiation, which it absorbs practically completely in wavelengths $\lambda < 2900$ Å, thereby reducing the solar heat flux by approximately 2%). But the aerosols, which are responsible for considerable scattering and absorption of radiation, are undoubtedly optically active as well. This means that the nonadiabatic heating ε are determined not only by the temperature and humidity fields (and also the turbulence field, which in turn is determined by these same fields and the wind field), but strictly speaking, also by the concentration fields of the various types of aerosol. However, the situation may not be so complex, since the characteristics of the aerosols are apparently to a large extent determined by the relative humidity of the air (i.e., by the ratio of the partial pressure of the water vapor e to its saturated value e_m at a given temperature), as Rozenberg[60] established from the results of measurements of the components of the complete matrix of light scattering in the atmosphere.

The conclusions reached from these measurements are enumerated below.

1. In a very dry atmosphere ($e/e_m < 0.3$) the aerosol does not contain water droplets but only the so-called Aitken nuclei or large ions (apparently of organic origin) with a discrete spectrum of sizes on the order of $0.01–0.1\,\mu$, in quantities on the order of thousands per cm^3, and also small numbers of salt particles with dimensions on the order of $0.1–1\,\mu$, the optical effect of which is very small (the visibility is greater than 40–50 km).

2. When $e/e_m \approx 0.3–0.4$, moisture begins to condense on the Aitken nuclei or the large ions. As the ratio e/e_m grows, the size of the particles (but not their number) increases, and their anisotropy decreases due to the growth of the water covering. When this happens, the dimensions of the largest particles correspond to the most optically active region from $0.1–0.3\,\mu$; a haze appears reducing the visibility to a little above 3 km (this phenomenon is observed in the atmosphere 90–95% of the time).

3. When $e/e_m \gtrsim 0.8$, the condensation apparently begins to involve particles, and the haze particles ($0.1–0.3\,\mu$) disappear. There is then a continuous size distribution

of particles with components of $1-5\,\mu$ (producing a rainbow) and $12-15\,\mu$ (which influence the halo portion of the scattering indicatrix) separated out. A foggy haze then appears with a visibility of $1-3$ km.

4. When $e/e_m \approx 1$ the particles of the foggy mist that have sizes of $1-5\,\mu$ grow to sizes of $8-12\,\mu$ (the component creating coronas and other features of the halo part of the scattering indicatrix). In the continuous spectrum of particle sizes another component from $18-25\,\mu$ separates; this fraction determines the brightness of scattered light in the halo region and the features of rainbows. Here the droplet sizes follow the temperature fluctuations, creating the wreath structure of fog with a visibility of less than 1 km.

17. The Regulating Role of Clouds

The primary source of the influx of energy into the atmosphere and ocean is the solar radiation flux to the earth. Careful measurements of this flux carried out to this day have revealed no noticeable time variations in it; for this reason, it has acquired the name *solar constant*. But if energy is supplied uniformly, then why does the A–AL system function nonuniformly? Why do long-term weather anomalies occur in it, so that the weather of a given year can turn out to be different from the weather of the preceding year? Clearly, if we cannot discover the source of this variability in external factors, we must make a search within the internal mechanisms that regulate the interactions among the various components of the system we are considering.

One of these regulators is immediately revealed if the following questions are answered: (1) What is it that most strongly influences the conditions of the passage of solar rays to the underlying surface (and thereby the conditions of heat supply to the active layer of the soil or the sea)? (2) What can intercept upward-moving heat radiation? (3) Where is heat given off when water vapor condenses? There is an obvious answer: the cloud cover. It is undoubtedly an effective regulator of the processes by which the constant flux of solar radiation is transformed into atmospheric heating that is unevenly distributed in space and time. It is especially important that the cloud cover is variable and that its evolution is determined by the same large-scale atmospheric motions whose energy supplies it regulates: the wind, which strongly influences the rate of evaporation of moisture from the underlying surface and which is responsible for the horizontal transfer of moisture, and the vertical motions that lead to the growth or disappearance of clouds. In other words, the cloud cover is a *regulator with feedback mechanisms*.[14, 61]

Cloudiness is probably not the only such regulator. For example, there are the ocean currents produced by the wind: They transfer heat and thereby influence the spatial distribution of heat flux from the ocean to the atmosphere; this distribution determines the energy supply of the wind field, so that there is a feedback in the mechanism of the generation of wind currents (we shall deal further with this question below). However, cloudiness is evidently the most effective regulator of the internal interactions in the A–AL system, so that one can assume that long-term weather anomalies on the earth are created mainly as a result of the *variable* cloudiness in the earth's atmosphere (according to this point of view, neither Mars with its cloudless atmosphere nor Venus with its continuous total cloudiness can have long-term weather anomalies such as those on Earth). Consequently, the cloud field must be included among the meteorological fields to be calculated in the theory of long-range weather forecasts.[14]

The regulating action of clouds is accurately described by the dynamic equations for the nonadiabatic processes, the equations presented in the two preceding sections. In a simplified form, this action can be represented as follows. At the initial instant of time, let there be an increased amount of heat in the ocean. The ocean will cause an intensified warming of the atmosphere, upward motions will develop in the atmosphere, and an increased amount of cloudiness will be produced. The cloudiness will cause an increased screening of the solar radiation, and a negative anomaly of the solar heat flux will appear. The ocean, obtaining less heat than it gives off to the atmosphere, will cool and begin to cool the atmosphere; downward motions will develop in the atmosphere, and the clouds will begin to disappear. With a decreased amount of clouds, the ocean will now undergo an intense warming, the conditions with which we began will be established, and the whole process will be repeated. Thus, owing to the presence of clouds, oscillations can occur in the A–AL system.

In order to estimate the possible periods of such oscillations, Gavrilin and Monin[62] * examined a simplified model of the A–AL system, for which the following hold:

1. The equation for the evolution of the potential vorticity (15.2) is governed by the quasi-geostrophic approximation and is linearized with respect to the rest state.

2. Only horizontal mixing is taken into account in calculating the heat flux due to the turbulent heat conductivity, i.e., it is assumed that $\varepsilon'_{turb}/c_p = K_h \nabla^2 T'$, where K_h is the coefficient of horizontal mixing.

3. It is assumed that in the absence of clouds (a probability of $\frac{1}{2}$ is assigned to this event) there are anomalies only of the long-wave radiant heat flux. The anomalies are described by Newton's law $\varepsilon'_r/c_p = -T'/\tau_r$, where τ_r is the "typical time of radiational smoothing of temperature inhomogeneities." It is also assumed that with clouds present (also with a probability of $\frac{1}{2}$) there are anomalies only of the heat flux due to phase transitions of the moisture. These anomalies are described by the linearized equation

$$\frac{\varepsilon'_\Phi}{c_p} \approx \left(1 - \frac{\alpha_0^2}{\alpha_b^2}\right)\left(\frac{\partial T'}{\partial t} - K_h \nabla^2 T'\right),$$

where α_0^2 is the static stability parameter, introduced in (7.2), and α_b^2 is the analogous parameter in the moist-adiabatic atmosphere; the average anomalies of the heat flux are then of the form $\varepsilon'_{turb} + \frac{1}{2}\varepsilon'_r + \frac{1}{2}\varepsilon'_\Phi$. The equation for the evolution of the potential vorticity (15.2) in this simplified model thus reduces to the form

$$\left(\frac{\partial}{\partial t} - K_h \nabla^2\right)\left(\alpha^2 L_0^2 \nabla^2 z' + \frac{\partial}{\partial p} p^2 \frac{\partial z'}{\partial p}\right) = -\frac{1}{\tau_0}\frac{\partial}{\partial p} p^2 \frac{\partial z'}{\partial p}, \tag{17.1}$$

where $1/\alpha^2 = \frac{1}{2}(1/\alpha_0^2 + 1/\alpha_b^2)$ and $\tau_0 = 2(\alpha_0^2/\alpha^2)\tau_r$. In order to evaluate the possible periods of oscillation, it is necessary to consider elementary wave solutions for $z'(x,y,p,t)$ of equation (17.1), whose functional dependence on (x,y,t) is expressed as $\exp\{i[(m_1 x + m_2 y)/L_0 - \omega t]\}$. From (17.1) it follows that the ampli-

*See also B. L. Gavrilin, NCAR manuscript No. 68–206, Sept. 1968.

tudes of such waves, possessing the proper regularity at $p \to 0$, will be related to p as p^{μ}, where

$$\mu = -\frac{1}{2} + \left[\frac{1}{4} + \alpha^2 m^2 \left(1 - \frac{1}{1 + (\tau_0/\tau_h)m^2 - i\tau_0\omega} \right) \right]^{1/2}, \tag{17.2}$$

with $m = (m_1{}^2 + m_2{}^2)^{1/2}$, the dimensionless horizontal wave number, and $\tau_h = L_0{}^2/K_h$, the "typical horizontal mixing time for the atmosphere," which apparently has a magnitude on the order of weeks. For oscillations with periods much larger than $(2\pi/m^2)\tau_h$, the dependence of μ on ω can obviously be neglected.

On the surface of the ocean ($p = p_0$), in the first place, boundary condition (15.3) must be satisfied for the function z'. In (15.3) we must set $w_0 = 0$ and $\varepsilon = \frac{1}{2}\varepsilon'_r + \frac{1}{2}\varepsilon'_\Phi$; moreover, Newton's law for the radiational heat exchange at the boundary is written in the form $\varepsilon'_r/c_p = -\frac{1}{2}(T' - T'_w)/\tau_r$, so that this first boundary condition reduces to the form

$$\frac{\partial}{\partial t}\left(p\frac{\partial z'}{\partial p} + \alpha^2 z' \right) = -\frac{1}{2\tau_0}\left(p\frac{\partial z'}{\partial p} + \frac{R}{g}T''_w \right). \tag{17.3}$$

Here $T''_w(x,y,t)$ is the temperature anomaly of the ocean surface; it turns out to be a new unknown. For its determination it is necessary to introduce a second boundary condition at $p = p_0$: that the algebraic sum of the anomalies of all vertical heat fluxes vanish at the surface of the ocean; this condition takes the form

$$c_w\rho_w K_w\frac{\partial T''_w}{\partial z} - \mathscr{L}\rho K_z\frac{\partial q'}{\partial z} + \rho H\varepsilon'_r + c_p\frac{H}{\tau_s}T_0 n' = 0. \tag{17.4}$$

Here the first term is the turbulent heat flux in the water (c_w, ρ_w, and K_w are the specific heat capacity, the density, and the mixing coefficient in the water); note that in this model, the analogous flux in the air is not taken into account. The second term is the vertical flux of latent heat in the air (q is the specific humidity, and K_z is the vertical mixing coefficient in the air). In order to avoid calculating the humidity field, in this model an additional simplification is made (somewhat underestimating the rate of evaporation):

4. Assuming that the specific humidity q is equal to the saturation value $q_m = (R/R_v)(e_m(T)/p)$ at the surface of the ocean, the value of $(\partial q/\partial z)_{z=0}$ can be replaced approximately by $\partial q_m/\partial z = -(\rho g/p)[(T/e_m)(\partial e_m/\partial T)(p/T)(\partial T/\partial p) - 1]q_m$. The anomalies of the latent heat flux then turn out to be proportional to $q'_m \approx (\partial q_m/\partial T)T'_w$ and can be written in the form $c_p\rho(H/\tau_\Phi)T''_w$, where τ_Φ is the "typical time for the vertical transfer of latent heat," and H is the altitude of the homogeneous atmosphere.

The third term in (17.4) is the anomaly of the long-wave radiative heat flux, which is determined as in (17.3). The fourth term is the anomaly of the solar heat flux; the complete expression for it is of the form $(1 - n)c_p\rho(H/\tau_s)\, T_0$, where $n = \frac{1}{2} + n'$ is the cloud index, T_0 is the average air temperature at sea level, and τ_s is the "typical time for solar heat input," defined by this equation. In order to avoid calculating the cloud field, the following simplification is made in the model:

5. It is assumed that the anomalies in the amount of cloudiness n' are proportional to the vertical velocity w at the level of maximum cloud occurrence $p = p_1$, i.e., $n' = (1/w_0{}^*)\,(\rho g w)_{p=p_1}$, where $w_0{}^*$ is an empirical coefficient (for which the value $w_0{}^* \approx 0.9$ mb/h is obtained from Lewis's previously mentioned nomogram[63]). The quantity $\tau_n = (p_0 - p_1)/w_0{}^*$ will represent the "typical time of cloud formation." Setting $\rho g w = \rho g (\partial z'/\partial t) - w^*$ and determining w^* from the thermodynamic energy equation, we get

$$\rho g w = \frac{g p}{\alpha_0{}^2 c_0{}^2}\left[\frac{\partial}{\partial t}\left(p\frac{\partial z'}{\partial p} + \alpha_0{}^2 z' \right) + \frac{R}{g}\frac{\varepsilon'}{c_p} \right]$$

where ε' is determined in the same way as for equation (17.1).

So far, in conditions (17.3) and (17.4) there is still one "extra" unknown: the value of $(\partial T_w'/\partial z)_{z=0}$. For its determination it is necessary to carry out a calculation of the water-temperature field in the ocean. For this purpose, in our model the following simplification is made:

6. Currents and horizontal mixing in the ocean are not taken into account, and for the calculation of the field T_w' the simple heat conduction equation

$$\frac{\partial T_w'}{\partial t} = K_w \frac{\partial^2 T_w'}{\partial z^2}$$

is used along with the condition of damping of temperature anomalies T_w' at the lower boundary of the active layer $z = -H_w$. It is then easy to verify that, for elementary wave solutions $T_w'(x,y,z,t)$ of the form indicated above, the relation

$$\frac{\partial T_w'}{\partial z} = \frac{T_w'}{H_w}\sqrt{i\tau_w\omega}\coth\sqrt{i\tau_w\omega} \quad \text{at} \quad z = 0 \tag{17.6}$$

will be satisfied, where $\tau_w = H_w^2/K_w$ is the "typical mixing time for the ocean." The value of τ_w is apparently very large, since, as a result of the strongly stable density stratification in the ocean, the mixing coefficient K_w is small (for example, values on the order of 10 cm^2/sec were obtained for K_w in calculations by Ozmidov and Popov[64] from the vertical distribution of Sr 90 in the Atlantic Ocean). For oscillations with periods much smaller than τ_w, $\tau_w\omega \gg 1$, and in (17.6) we can set $\coth(i\tau_w\omega)^{1/2} \approx 1$.

Using the law $z' \sim p^\mu$ and relation (17.6), we can obtain a relation between the frequencies ω and the horizontal wave numbers m of the elementary oscillations; this relation can be used as a condition of the existence of nonzero solutions for z' and T_w'' from equations (17.3) and (17.4). For frequencies in the interval $m^2/\tau_h \gg \omega \gg 1/\tau_w$ this relation takes the form

$$(1 - 2i\Omega)(\sqrt{i\Omega} - \xi) = \eta, \tag{17.7}$$

where $\Omega = \tau_0\omega[1 + (\alpha^2/\mu)]$ is the dimensionless frequency,

$$\xi = \lambda\left(\zeta - \frac{\tau_0}{\tau_\Phi} - \frac{\alpha_0{}^2}{\alpha^2}\right)$$

and

$$\eta = \lambda\left[\zeta\left(1 + 2\frac{\tau_0}{\tau_h}m^2\right) + \frac{\alpha_0{}^2}{\alpha^2}\right],$$

with

$$\lambda = \frac{c_p\rho H}{c_w\rho_w H_w}\sqrt{\frac{\tau_w}{\tau_0}\left(1 + \frac{\alpha^2}{\mu}\right)}$$

and

$$\zeta = \frac{(p_1/p_0)^\mu}{2(p_0/p_1 - 1)\alpha^2}\frac{\tau_n}{\tau_s}.$$

If $2(\tau_0/\tau_h)m^2 \gg 1$ and $\eta \gg 1$, then from (17.7) we get $\Omega \approx \frac{1}{2}(\eta/2)^{2/3}(i \pm 3^{1/2})$. This corresponds, in the first place, to growing oscillations, so that the presence of a feedback mechanism creates a swinging in the system (probably limited to the action of the nonlinear terms that the model does not take into account). Second, for the periods τ of such oscillations, the equation

$$\tau = C\left(\frac{\tau_0\tau_h\tau_s}{\tau_n\sqrt{\tau_w}}\right)^{2/3}\left(\frac{L}{L_0}\right)^{4/3} \tag{17.8}$$

is obtained, where L is the wavelength, and

$$C = \frac{2}{\sqrt{3}}\sqrt[3]{\frac{2}{\pi}}\left[\frac{\alpha^2(1 + \alpha^2/\mu)(p_0/p_1 - 1)}{(p_1/p_0)^\mu}\frac{c_w\rho_w H_w}{c_p\rho H}\right]^{2/3} \approx 1.$$

For wavelengths on the order of L_0 and reasonable values of the parameters that occur in (17.8), periods τ are obtained on the order of months. Thus, according to this model, the interactions between the ocean and the atmosphere that are regulated by cloudiness can have the character of oscillations with periods on the order of months.

18. Numerical Experiments

The full set of dynamic equations of a given physical system presented in one of the approximate forms, along with the corresponding boundary conditions and with the *algorithm for the numerical solution* of these equations—inevitably containing means from a finite-difference approximation of the continuous fields describing the system—form a physicomathematical *model* of the system. The algorithm for the numerical solution of the equations is an important part of the model, since using an algorithm results in a finite-dimensional system that, alas, not infrequently is far from being equivalent to the original continuous system.

Solving the equations of the model for some assigned values of the "external parameters" can be called a *numerical experiment* in the dynamics of the physical system in question. Thus, finding a stationary solution (or a solution with a one-year period) for the equations of a model of the planet's atmosphere (or better, the A–AL system) is a *numerical experiment in the climatic background* (see the examples of such stationary solutions by Blinova and others[2-6] mentioned in section 13). Finding a nonstationary solution for concrete initial data that fix some actual instantaneous state of the system is a *numerical experiment in forecasting*. Numerical experiments on the short-range prediction of meteorological fields (with the adiabatic approximation) were begun in Kibel's laboratory as early as 1940. Starting in 1950, they were carried out in several countries and were based on equation (7.6′), then on (7.4) and (8.2), (8.3), and finally, on the "primitive equations."

Numerical experiments in the global prediction of meteorological fields were conducted by Blinova, beginning in 1943, first on the basis of the simplest dynamic equation (8.2′) and the heat transfer equation

$$\left(\frac{\partial}{\partial t} + \alpha \frac{\partial}{\partial \lambda}\right) T' - \frac{2M \cos \theta}{a^2} \frac{\partial \psi}{\partial \lambda} = K_h \nabla^2 T', \qquad (18.1)$$

linearized in a similar way, where M is the meridional temperature difference (the horizontal turbulent heat conductivity was introduced here to prevent the growth of small-scale inhomogeneities). The equations were then complemented by the prediction (allowing for nonlinearities) of the circulation index α; baroclinic models began to be applied; moreover, the consideration of the vertical heat conductivity in (18.1) required the use of a boundary condition similar to (17.4) and consideration of the temperature in the soil or the ocean, i.e., construction of a nonadiabatic model. The improvements enumerated are summarized in a review by Blinova;[7] her papers of recent years[8-12] have been devoted to the development of non-adiabatic models.*

Integrating a model's equations over long time intervals (several months or even several years) with idealized initial data corresponding, e.g., to a slightly perturbed rest state is a *numerical experiment on the statistical regime* of the system under investigation. Such numerical experiments in the *general circulation of the atmosphere* became popular after the success of the first such experiment, which Phillips[65] carried out in 1956. The most detailed experiments have been conducted by Smagorinsky,[66-72] Mintz[73-76] (see also the associated paper by Mesinger[77]) and Leith;[78-81] experiments by Matsumoto[82] and Chen,[83-84] papers by Kasahara and Washington[85] and Gambo,[86] and experiments by Adem[87-91] should also be mentioned. A review of the principal experiments on the general

---

*See also the papers by Blinova in *Tr. Gidromettsentra SSSR* [Trans. Hydrometeorological Center USSR], No. 15, 3 and 26 (1967) and No. 31, 3 (1968); the latter paper contains the exact solution for the equations of the nonlinear two-level model.

atmospheric circulation has been published by Gavrilin[92] (it has also appeared in English in the 1967 GARP conference report[13]).

Numerical experiments on the circulation of the ocean have been conducted by Sarkisyan[93-95] and Garmatyuk and Sarkisyan[96] (see also Sarkisyan's book[97]), Bryan,[98] and Bryan and Cox.[99] Finally, Manabe and Bryan carried out a numerical experiment on the general circulation of the atmosphere and ocean in their interaction; preliminary results were reported by Manabe at the Fourteenth General Assembly of the International Geodetic and Geophysical Union in 1967 in Switzerland.* Below is a more detailed discussion of the numerical experiments on the general circulation of the atmosphere by Phillips, Smagorinsky, Mintz, and Leith.

Phillips's calculation was carried out according to a two-level quasi-geostrophic model on a grid in the β plane with intervals $\Delta x = 375$ km and $\Delta y = 625$ km, and the instantaneous state of the atmosphere was characterized by 450 numbers; the time interval was 0.5–2 hours, and in the main experiment the equations of the model were integrated for a 31-day period. The nonadiabatic factors the model took into account were the horizontal turbulent viscosity and the heat conductivity (with the mixing coefficient $K_h = 10^5$ m^2/sec), the surface friction (with the resistance coefficient $c_f = \tau/\rho U^2 = 3 \times 10^{-3}$, where τ is the frictional stress and U is the wind velocity at the earth's surface), and the radiative heating, which was given by the empirical formula $\varepsilon_r = -2H_0 y/W$ (where y is the coordinate along the meridian, $W = 5000$ km, and $H_0 = 20$ cm^2/sec^3); the condensational heating was taken into account only by reducing the static stability parameter by 20%.

The experiments by Smagorinsky, Mintz, and Leith were calculated with the aid of the primitive equations on a sphere. Smagorinsky[66] used a two-level model (with the additional requirement that the velocity divergence averaged over altitude vanish, which allows the two-dimensional gravity waves to be filtered out of the solutions) with intervals of 5° in latitude and longitude. The calculation was performed only for the zone between the equator and 64° N; the quantity of numbers characterizing the instantaneous state of the atmosphere was an order of magnitude greater than that used by Phillips. The time interval was 20 min, and

*See S. Manabe and K. Bryan, *Monthly Weather Rev.* **97**, 739–829 (1969); among other numerical experiments and studies recently carried out by them, there are: W. Washington, *Sci. J.* **4**, No. 11, 34–41 (1968); K. Bryan and M. D. Cox, *J. Atmos. Sci.* **25**, 945–978 (1968); K. Miyakoda and staff members, *Proc. Symp. Numerical Weather Prediction, Tokyo, 1969*, pp. 1-6 (4); A. Arakawa, A. Katayama, and Y. Mintz, ibid., p. 7 (4).

the main experiment was computed to span a 60-day period. The description of nonadiabatic factors differed from Phillips's in that a more detailed account was taken of the turbulent exchange (with the coefficients of horizontal and vertical viscosity depending on the velocity field) and in that the dependence of the radiative heat influx on the temperature, $\varepsilon_r = f(y) - C(\Phi_1 - \Phi_3)/T_m$ (where Φ_1 and Φ_3 are the geopotential at two levels, T_m is the temperature of radiative equilibrium, and C is an empirical constant), was taken into account.

Mintz used a two-level model with a horizontal grid of 1000 points on a sphere. He allowed for the influence of large mountain masses and the differences in the thermophysical properties of the oceans and the continents: He assumed that the oceans have an infinite heat capacity (and the temperature of their surfaces was specified) and that the continents have zero heat conductivity. The time interval was 12 min, and the experiment was computed for a one-year span. The description of nonadiabatic factors was complemented by a consideration of the vertical transfer of heat by convection (proportional to the difference between the actual and the moist-adiabatic temperature lapse rates). The values of ε_{rL} were chosen in the form of a function of the temperature and altitude; this function was obtained in the paper by Takahashi, Katayama, and Asakura[100] in a calculation of the mean climatic distribution of humidity and cloudiness. Finally, in calculating ε_{rS} it was assumed that the solar heat flux, reduced because of the planetary albedo (whose dependence on latitude was obtained from *Tiros 4* data), is absorbed in the following proportions: 73% by the earth's surface, 15% by the lower layer of the atmosphere, and 12% by the upper layer.

Leith's experiment consisted of a calculation according to a six-level model with a horizontal grid of 2000 points on a sphere. He forecasted the surface pressure field and the three-dimensional fields of the temperature, the horizontal wind-velocity components, and the humidity (incidental computations were made of the horizontal velocity divergence D, the individual pressure change w^*, and the geopotential gz). The quantity of numbers characterizing the instantaneous state of the atmosphere was here two orders of magnitude greater than that used by Phillips. Consideration was given to the difference between the oceans and the continents (but not to the topography of the continents). The time interval was 20 min, and the experiment was computed for a one-year span. In taking the nonadiabatic factors into account, the values of ε_{rL} and ε_{rS} were chosen, respectively, in the form of empirical functions of the altitude and the mass of water vapor along the ray paths; heating due to the condensation of moisture and the falling of precipitation were taken into account.

Finally, in the experiments by Smagorinsky and his coauthors,[68, 70] a nine-level model was used with a horizontal grid of 1250 points on the Northern Hemisphere (a rectangular grid on a map of a stereographic projection with $N = 20$ grid points between the equator and the pole). The time interval was 10 min, and the experiment described in the paper by Smagorinsky, Manabe, and Holloway[68] was computed for a 300-day span and the one described in the paper by Manabe,

Smagorinsky, and Strickler[70] for a 187-day span (in an article by Smagorinsky et al.[71] experiments in forecasting are described with the same model for $N = 40$, a time interval of 5 min, and time spans up to 108 h). Of the nonadiabatic factors, the influence of turbulence was described as in Smagorinsky's first model,[66] radiative heat influxes ε_r were calculated according to Manabe and Strickler's[50] scheme (similar to the one presented in section 16), using the average climatic vertical profiles of the water vapor, CO_2, ozone, and cloud cover; in writing a boundary condition like (17.4) for the surface of the earth, the heat capacity of the earth is assumed to be zero. In reference 68 a "dry" model was used: The humidity field was not taken into account, and in order to make allowance for the stabilizing influence of moisture convection, the vertical temperature gradients, which exceed the moist-adiabatic gradient, were automatically replaced by the moist-adiabatic gradient. In reference 70, on the other hand, a "wet" model was used in which the humidity field was considered; the effect of dry and wet convection was taken into account as in reference 68, and it was assumed that all condensed moisture immediately falls out in the form of precipitation.

The primary success of these numerical experiments consists in their reproduction of the fluctuating character of the general circulation of the atmosphere with quite realistic amplitudes for the synoptic fluctuations of the meteorological fields. Thus, in the first experiment by Phillips, the atmosphere, starting from a state of rest, evolved to a state with wind velocities on the order of 10 m/sec and pressure-pulsation amplitudes on the order of 20 mb. By way of illustration, Fig. 25 shows the time changes in the average kinetic

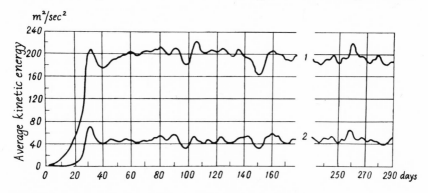

Fig. 25 Time variations of the average kinetic energy of the entire atmosphere E/M (with an initial state of rest) in Mintz's experiment:[73] (1) the total kinetic energy; (2) the kinetic energy of meridional motions.

energy of the entire atmosphere E/M in Mintz's experiment. The chart shows that, after an initial rest state, the energy of the atmospheric motions rapidly increases (as a result of nonadiabatic heating), and after 30–40 days a quasi-stationary fluctuating regime is established. The first part of the experiment, which corresponds to the left half of the graph, was computed without taking mountain masses into account, and the second part, the right side of the graph, did allow for the influence of mountains; this influence had practically no effect on the total kinetic energy.

Some fine features of the synoptic fluctuations have also been successfully reproduced in the experiments. In Smagorinsky's experiment described in reference 66 a spectral selectivity of baroclinic instability was discovered: the predominant growth of waves with numbers $m = 5$ or 6 (m is the number of waves on a circle of latitude) —and global oscillations with a period on the order of two weeks that is called the "index cycle" appeared; see section 3.4 and Fig. 4.

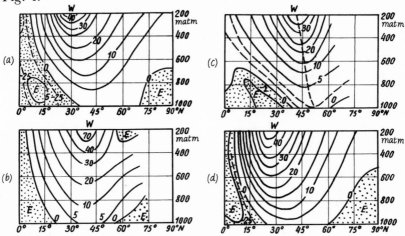

Fig. 26 Distribution of the average zonal wind (in m/sec) in the Northern Hemisphere: (a) real data for the winter season, after Mintz;[101] (b) the average for the 8th day to the 30th day according to Phillips's experiment;[65] (c) the average for the 17th day to the 39th day according to Smagorinsky's experiment;[66] (d) the average for the 256th day to the 285th day according to Mintz's experiment.[73]

Fig. 27 A map of the average atmospheric pressure at sea level: (a) the average for the 256th day to the 285th day according to Mintz's experiment;[73] (b) real data for January.

Numerical experiments have also been successful in reproducing several characteristics of the "climatic background," i.e., the meteorological fields averaged over long periods of time. Thus, Phillips succeeded in reproducing the triple-cell structure of the zonal circulation with easterly surface winds (the trade winds) in the tropics, a west-east flow in the middle latitudes with a maximum (the jet stream) in the upper troposphere, and westerly surface winds in the polar zone; see Fig. 26, in which this finding by Phillips is compared with similar results obtained by Smagorinsky[66] and Mintz[101] and with actual data (taken from Mintz's paper[101]). Figure 27 gives a comparison between actual data and a map of the average atmospheric pressure at sea level obtained in Mintz's experiment; the agreement between the two maps can be considered satisfactory (except for a few details: an extra maximum in Greenland and an extra minimum in North America were obtained in the experi-

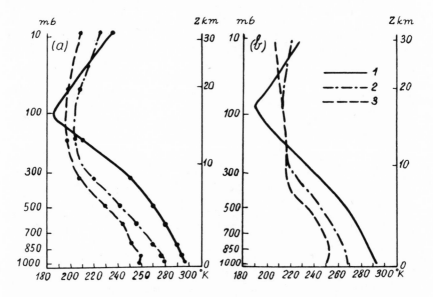

Fig. 28 Vertical profiles of the average annual zonal temperature: (1) at 10°N; (2) at 50°N; (3) at 90°N. Profile (a) is from the experiment by Smagorinsky et al.[68] with a "dry" model; profile (b) is according to real data.

ment, but the experiment did not obtain the maximum in the north-eastern Pacific, the minima over South Africa and South America, and the weak subtropical maxima in the Southern Hemisphere).

Figure 28 compares actual data and the vertical profiles of the average annual temperature obtained in the experiment by Smagorinsky and his coauthors[68] with a "dry" model. According to this experiment, the difference of the altitudes of the tropopause at the equator and at the pole is equal to 7 km as against the actual 10 km; the difference of the corresponding surface temperatures (42°) is close to the actual difference, but the meridional temperature gradient in the upper troposphere of the middle latitudes turned out to be too large. Figure 29 gives a comparison of an actual

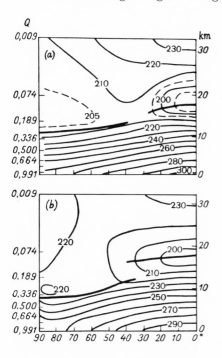

Fig. 29 Meridional section of the average annual zonal temperature field: (a) according to the experiment by Smagorinsky et al.[70] with a "wet" model; profile (b) is according to real data.

meridional section of the average annual zonal temperature with that obtained in the experiment by Smagorinsky et al.[70] with a wet model.

The achievements reflected in Figs. 26–29 are not in themselves new: On the whole, they repeat the results of numerical experiments that were carried out much earlier on the climatic background (Blinova,[2, 3] Mashkovich,[4] and others) and first gave a theoretical explanation of the corresponding features of the general circulation of the atmosphere. What is new is that now they are obtained, not with the aid of the stationary solutions of the dynamic equations, but by an averaging of the nonstationary solutions, so that they characterize the advantages of models designed to describe the *fluctuations* of the general circulation of the atmosphere.

From their experimental data, Smagorinsky[66, 68] and Gambo[86] computed the rates of the various energy transformations in the atmosphere. The results of these computations appear in Fig. 30 in comparison with empirical estimates by Oort[102] (actual data can also be found in the papers by White and Saltzman,[103] Krueger, Winston, and Haynes,[104] and Wiin-Nielsen et al.[105-109]). From this chart it is clear that numerical experiments in general can already duplicate the energetics of the atmosphere fairly well, even though there are some discrepancies with empirical estimates (an overstating of the ratios of the zonal to nonzonal components of the potential energy and kinetic energy $\bar{\mathscr{P}}/\mathscr{P}'$ and $\bar{\mathscr{K}}/\mathscr{K}'$ and the rate of the transformation $\mathscr{K}' \rightarrow \bar{\mathscr{K}}$ and an overstating of the generation rate of potential energy $\bar{\mathscr{P}}$).

In the experiments by Smagorinsky et al.[68, 70] the flux fields of the momentum, heat, and humidity were also calculated in detail and compared with the available empirical estimates (it should be noted that actual data on the *climatic character* of these fields is virtually unavailable; several estimates, claimed to be climatic by their authors, were obtained, not at all from measurement data, but from crude calculation methods that are much less reliable than numerical experiments).

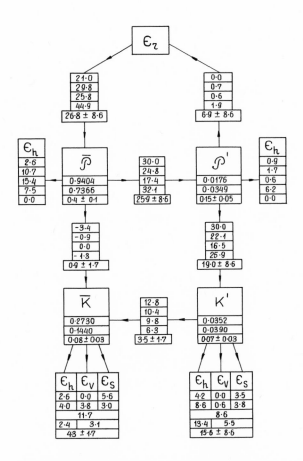

Fig. 30 Energy conversions in the atmosphere on an average over the Northern Hemisphere: ε_r is the external energy source; $\overline{\mathscr{P}}$ and $\mathscr{P}'$ are the zonal and nonzonal components of the potential energy and $\overline{\mathscr{K}}$ and $\mathscr{K}'$ are the zonal and nonzonal components of the kinetic energy; ε_h, ε_v, and ε_s are the rates of energy dissipation due to the horizontal and vertical turbulent mixing and the friction at the surface of the earth. The energy values (in J/g) are given from top to bottom according to experiments by Smagorinsky[66, 56] and Oort's real data;[102] the rates of energy conversion (in 10^{-3} J/g-day, positive with respect to the direction of the arrows) are given from top to bottom according to the experiments by Phillips, Smagorinsky,[66] Gambo,[86] and Smagorinsky[68] and from Oort's data.

Mintz[73] conducted some interesting experiments with the aid of his model. As it turned out, when the Himalayas are removed, the Siberian maximum does not appear on the chart of the average pressure at sea level; erecting a smooth wall at the equator causes the easterly winds above the equator to disappear; when the daily variation of solar radiation is taken into account, semidiurnal oscillations of the surface pressure appear, with an amplitude of about 1 mb at the equator and with a phase close to that which is observed. With the aid of Mintz's model, Mesinger[77] calculated the 30-day trajectory of counterbalanced sounding balloons at levels of 800 and 300 mb; it turned out that the balloons, initially distributed uniformly in the atmosphere, quickly abandon the tropical zone and display a tendency to accumulate in regions of reduced pressure in the middle latitudes. With the aid of Smagorinsky's, Mintz's, and Leith's models, Charney[15] made estimates of the periods of predictability for the synoptic processes (they can evidently serve as lower-limit estimates for the real atmosphere); these estimates will be discussed further in section 20.

The numerical experiments that have been carried out enabled Smagorinsky (see the 1967 GARP conference report[13]) to estimate the typical reaction times of the atmosphere under various non-adiabatic influences: These are one day for the heat released by the condensation of moisture ε_Φ and for the viscous dissipation of energy, three to four days for turbulent fluxes of heat and latent heat, and one week for radiative heat fluxes; the reaction of the active layer of the ocean becomes appreciable within periods of one to two weeks.

Numerical experiments make it possible to ascertain the roles of various factors in the formation and change of the climate. Consequently, they constitute a method for explaining the features of our modern climate as well as for making paleoclimatic reconstructions and predicting future changes in the climate, including those resulting from man's influence.

How would the climate change if there were changes in the solar constant? Will the increasing production of carbon dioxide due to the burning of fuel lead to its accumulation in the atmosphere, causing a warming as a result of the accompanying amplification of the greenhouse effect (and if so, what kind of warming will it be)? Or will there only be an increase in the absorption of CO_2 by the oceans and the photosynthesizing plants? What could be the result of the melting of the arctic ice fields: their restoration (which has apparently taken place already after the "climatic optimum" of the fortieth to twentieth centuries B.C. and the Viking era of the eighth to tenth centuries A.D.; see section 3.8) or an irreversible global warming? These and similar questions have already been repeatedly and most emphatically answered both by conscientious people that have erred and by outspoken lovers of public sensation (for the latter, unreal problems such as the "thermal reclamation of the Arctic" are opportune, and not problems of economic significance —such as, for example, reversing the flow of the Ob River into central Asia—the solution of which imposes responsibility). But only numerical experiments with models that have been developed in sufficient detail constitute a method for the *scientific* solution of such problems.

For developing the models of the atmosphere (and then the A–AL system as well) in greater detail, for specifying the values of their parameters, and for checking the results of numerical experiments, it is necessary to obtain data from the global observation of the atmosphere and ocean. On the other hand, numerical experiments will probably be the primary method for utilizing the global data, so that such data will be fully exploitable only by laboratories that possess models of the atmosphere and are ready to carry out numerical experiments. The perfection of today's models and the acquisition of ever more complete real data on the state of the atmosphere and ocean should lead to the formulation of definitive models and to the selection of the initial conditions necessary for regular long-range weather predictions.

One may expect that in future models the quantity of numbers describing the instantaneous state of the atmosphere will grow by another 1.5 to 2 orders of magnitude: If the distances between the points on a horizontal grid are 110 km, there will be 4×10^4 such points, so that data on 4 fields at 20 levels will contain $4 \times 20 (4 \times 10^4) \approx 3 \times 10^6$ numbers. According to Smagorinsky's estimation (in the 1967 GARP conference report[13]), integrating the equations of a model for one hour ahead requires 24 iterations, so that in integrating for one day it will be necessary to compute $24 \times 24 (3 \times 10^6) \approx 2 \times 10^9$ new dependent variables to define the evolving meteorological fields.

If 500 arithmetic operations are carried out in computing a single dependent variable, then an experiment requiring the integration of the equations for a span of 500 days will require $500 \times 500 (2 \times 10^9) \approx 5 \times 10^{14}$ operations. If it is required that a forecast for 1 day take no more than half an hour to compute, so that an experiment with integration spanning 500 days will take 250 hours of machine time, then such an experiment will require a computing machine with a high-speed action of $(5 \times 10^{14})/(250 \times 60 \times 60) \approx 5 \times 10^8$ operations per second. This calculation is useful as a guide; the reader will here be able to change the specified parameters that he does not like in order to try to obtain results that seem more acceptable to him.

With qualitative regard to today's models, it is necessary first of all to allow for consideration of the variable cloudiness and its influence on the radiant energy flux (the importance of which was explained in section 17) and also to give a detailed account of the interaction between the atmosphere and the active layer of the ocean, to the consideration of which we now turn.

19. The Interaction of the Oceans and the Atmosphere

The important role of the interaction between the oceans and the atmosphere in the formation of large-scale atmospheric processes was long ago brought to light by several authors. Shuleykin showed that these interactions can lead to the occurrence of *self-oscillations* in the A–AL system; for example, there could occur relatively short-period (with periods on the order of one week) oscillations in the atmospheric temperature and pressure fields, reminiscent of standing waves in lakes (seiches), but with nodal lines revolving around certain centers on the coasts of the continents, as well as long-period (with periods of about 3.5 years) temperature oscillations in the northern branch of the Gulf Stream. It should also be noted that the presence of closed circulations of water in the oceans can lead to the separation in the oscillation spectrum of periods equal to the least common multiple of the circulation periods and one year.

Many authors have discovered a connection between the atmospheric processes and the temperature of the surface of the oceans T_w. For instance, Perloth[110] points out a relationship between T_w and the pressure at the center of Hurricane Esther of 1961; Pyke[111] explains the rapid development of cyclones in the winter near the eastern shores of the continents by the influence of the warm ocean surface; Bradbury[112] makes recommendations for using the field T_w in synoptic analysis. Useful studies were carried out by Semenov and Shushevskaya;[113, 114] from data for 1899–1939, they revealed the presence of a connection in the winter seasons between signs of anomalies of T_w in the North Atlantic and the recurrence of cyclones and also a connection between T_w in the North Atlantic and the temperature in Europe; these studies were, unfortunately, abandoned without reason.

In thorough investigations by Scherhag,[115, 116] close connections are shown to exist among the average annual fields of T_w, the surface pressure, and the precipitation. The influence of long-lasting anomalies of the field T_w (and the snow cover) on the character of the atmospheric circulation has been demonstrated for several specific instances in papers by Namias.[117-120] Sawyer[121] made a quantitative evaluation of this influence, based on data from empirical studies (by Arkhipova[122] and Schellard[123]) of the anomalies of the heat exchange between the North Atlantic and the atmosphere: Anomalies of the field of heat flux from the ocean to the atmosphere reach values of 80 cal/cm²-day and can exist for longer than a month (two-thirds of the magnitude of these anomalies is produced by the anomalies of the evaporation). Finally, Adem[90] conducted a numerical experiment on the long-range forecasting of the air-temperature field in the Northern Hemisphere for the winter of 1963, showing that the temperature anomalies of that season were determined by the initial anomalies of the field T_w (and the snow cover).

Especially interesting results in unmasking the processes of the large-scale interaction between the ocean and the atmosphere were recently obtained by Bjerknes.[124-128] In the first two of his papers, studying data on the North Atlantic led Bjerknes to assume the existence of a positive feedback between (1) the intensification of atmospheric circulation in the middle latitudes when the heat transfer from the ocean to the atmosphere in the trade-wind zone increases and (2) the intensification of the trade winds and the increase of the heat transfer there during the intensification of the circulation in the middle latitudes.

In two of Bjerknes's papers[126, 127] it was established that, during the winter of 1940–1941 and especially during the winter of 1957–1958 (and according to later studies, also in 1963–1964 and 1965–1966), the easterly winds in the eastern part of the equatorial zone of the Pacific were abnormally weak; a weakening was thereby produced in the rise of cold, deep water, and as a result there was a positive anomaly of T_w of up to 2–3°C (the so-called El Niño condition). The intensified heating of the atmosphere in this zone led to the intensification of the trade (meridional) circulation and therefore (in accordance with the conclusions of Bjerknes's earlier papers[124, 125]) to an intensification of the west-east flow in the middle latitudes of the Pacific Ocean and to a deepening of the pressure minimum in its northeastern part; see in Fig. 31 a comparison of maps of the atmospheric pressure at sea level for the normal winter of 1955–1956 and during the El Niño in 1957–1958. At the same time, in the Atlantic there was, on the contrary, a weakening of the west-east flow in the middle latitudes and a filling of the pressure minimum over Iceland; there was therefore a weakening of the easterly winds north of Iceland, as a result of which the Arctic was under the influence of an anticyclone north of Alaska. Thus the eastern part of the equatorial zone of the Pacific Ocean turned out to be the "weather kitchen" for the entire Northern Hemisphere.

Bjerknes[128] also studied the much longer processes of ocean-atmosphere interaction that took place during the "Little Ice Age" in the seventeenth to nineteenth centuries A.D. (manifestations of this period in data on the altitude of the firn line and the length of the Norwegian and Icelandic glaciers are given in Fig. 32, after

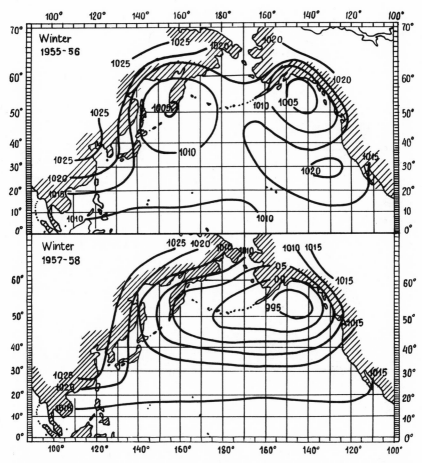

Fig. 31   Maps of the atmospheric pressure at sea level during the winter season (December to February) in the northern part of the Pacific Ocean: (a) in the normal winter of 1955–1956; (b) in the winter of 1957–1958 during a period of maximum warming (El Niño). After Bjerknes.[126]

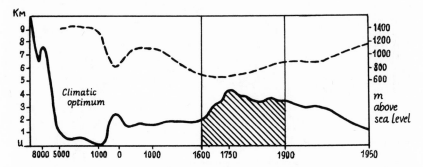

Fig. 32   Variations of the length of Norwegian and Icelandic glaciers in km (the solid line) and the altitudes of their bedding in m above sea level (the dotted line). After Ahlmann.[129]

Fig. 33   Temperature anomalies of the water surface in the North Atlantic during 1780–1820 relative to the average for 1887–1899 and 1921–1938. After Lamb and Johnson[130] (dotted line after Bjerknes[128]).

Ahlmann[129]). The temperature drop during this period cannot be explained by changes of the configuration of the continents and the altitudes of mountains: They undoubtedly were the same as they are now. However, as Bjerknes established, during this period the winters were characterized by a weakened circulation that led to a positive anomaly of T_w in the Atlantic in the region of the Sargasso Sea and a negative one in the area of Iceland (see Fig. 33, adopted from Lamb and Johnson[130]). In the presence of the positive feedback referred to above, these anomalies of T_w aggravated the subsequent weakening of the atmospheric circulation, and possibly only the increased northward transfer of heat by the circulation of the waters of the Atlantic altered the dangerous direction of the climatic trend during this period.

The last segments of both curves in Fig. 32 deserve special attention: They reflect the global warming following the end of the "Little Ice Age" that has literally taken place before our eyes. This heating is shown in more detail by the temperature curves in Fig. 34, adopted from Mitchell's paper.[131] The geographic inhomogeneity of this warming, illustrated in the chart of Fig. 35 (after Mitchell[132]), is highly significant: The abrupt *warming of the Arctic winters* was accompanied by a slight *temperature drop on the continents* and in the tropical regions of the Atlantic and Indian oceans; the difference in the influence of the oceans and the continents was striking. After a logical analysis of the conceivable reasons for the changes in the climate (Fig. 36), Mitchell[133] suggested a stochastic model of the ocean-atmosphere interaction to explain the indicated warming.

Explaining the physical mechanism of the climatic warming of the first half of the twentieth century is undoubtedly one of the most important concrete problems of modern climatology. For its solution and for many other purposes, it is necessary to have estimates of the climatic average values of the meteorological fields, including the field of the heat fluxes of various kinds, with errors that are small compared to the typical amplitudes of the intrasecular oscillations of these fields (Kolesnikova and Monin;[134] this accuracy criterion

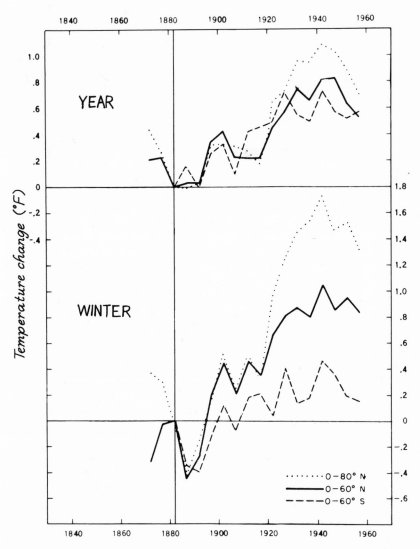

Fig. 34   Deviations of the five-year temperature averages in selected latitudinal zones from their average values for the five-year period of 1880–1884. After Mitchell.[131]

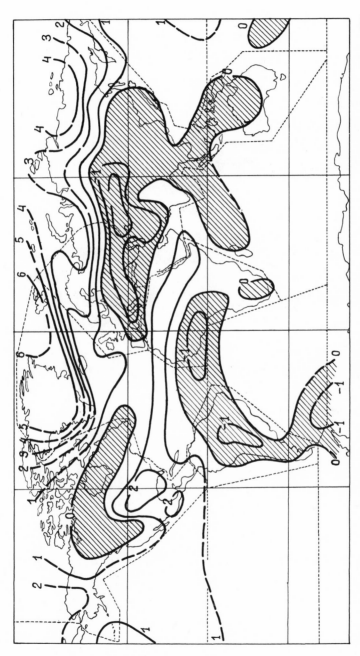

Fig. 35 A map of the changes in the average winter temperatures for 1920–1930 relative to 1900–1919. After Mitchell.[132] The isolines are in gradations of 1°F; regions in which cooling occurred are shaded.

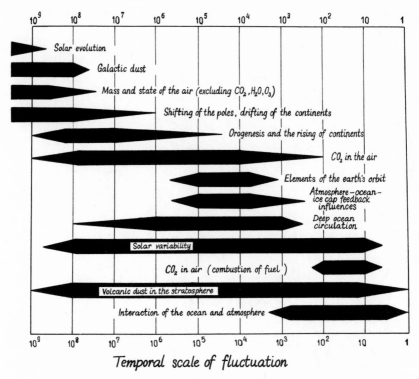

Temporal scale of fluctuation

Fig. 36 Logically conceivable reasons for changes in the climate and the corresponding typical times. After Mitchell.[133]

in modern climatology stands in opposition to the retrograde tendencies of the so-called balance study, in which the estimation of climatic average values is permitted from the data from only a few years and even a single year).

20. Predictability

In many areas of meteorology the concept of large-scale and small-scale motions is of great benefit, with the former described individually and the action of the latter considered only statistically. For example, in describing the distribution in the air of smoke emitted from a factory chimney, the large-scale air motion—the wind—is assigned individually, and the small-scale turbulence, which produces irregular curls in the smoke stream, is considered only statistically (by introducing exchange coefficients, let us say).

Of course, long-range weather forecasting is also related to such areas of meteorology. We want to be able to make long-range predictions of large-scale weather characteristics individually, but it is clearly impossible in practice (and unnecessary as well) to make individual long-range predictions of small-scale motions, e.g., those producing the curls in streams of smoke. Thus there arises the natural question (discussed in Obukhov's report[135]), What must be the dividing line between large-scale processes ("the weather"), which are forecasted individually, and small-scale processes ("turbulence"), which are described only statistically?

This division can be different for different forecast periods. In fact, for the shortest periods (up to twenty-four hours, say) we try to *individually* observe even the mesoscale processes [by using data from a grid with reduced steps between weather stations on synoptic charts called *"kol'tsovkami"* ("circlings")]; on the other hand, for generalizations over many years, it is advisable to describe all the synoptic processes *statistically* (by considering them, according to Defant's idea,[136] to be *macroturbulences*).

The actual unpredictability of small-scale motions over long periods requires some clarification. If we could precisely fix the initial state of all the small-scale motions and obtain the exact solutions for their dynamic equations, then, to put it abstractly, the periods of predictability would be in no way limited. Even the continuity of the hydrodynamic fields would not be an obstacle in theory: As continuous fields change linearly over very small distances, it would be sufficient to fix the initial values of the hydrodynamic fields only at the points of a spatial grid with distances smaller than the internal scale of turbulence $\lambda \sim (v^3/\varepsilon)^{1/4}$ (which has values on the order of millimeters in the atmosphere).

But we can fix the initial values of the meteorological fields only at points of a much coarser grid (with a horizontal interval 10^7 to

10^8 times larger than the internal scale λ), so that the individual motions with scales smaller than the interval are obviously not fixed at all; in addition, we make random errors in measuring, interpolating, and smoothing. Because of these initial errors, even with an exact solution of the exact dynamic equations, a forecast inevitably contains errors that, generally speaking, will be larger the longer the period of forecast (moreover, we use only approximate dynamic equations, and we approximate them for the purpose of numerical solution, e.g., by difference equations).

It is clear that forecasting individual processes will give additional information beyond that given by their statistical (climatic) description only as long as the errors in prediction do not exceed the average climatic variations of the quantities that are being predicted. The corresponding period of time can be called the *limit of predictability* of the processes under consideration. It obviously depends on (1) the type of processes (in particular, their scales) and the form of their quantitative characteristics that are considered, (2) the size and nature of the initial errors, and (3) the quality of the prognostic method. The task of determining the limits of predictability can be called the *predictability problem*.

It is possible that small changes of the initial state of the atmosphere (initial errors) can lead over long periods to important changes in the final state of the atmosphere (and create the predictability problem). This possibility was pointed out 20–25 years ago in a series of lectures by Kolmogorov in the following expressive form: Imagine two identical planets with perfectly identical atmospheric states. If a handkerchief is waved on one of them but not on the other, then how long will it take for the weather on these planets to become completely different?

The general mathematical formulation of the predictability problem (presented in one of my papers[137]) consists in the following: Let $\omega = \Sigma_L \omega_L$ be the states of the atmosphere represented in the form of a superposition of components with scales L (for example, the state of a quasi-geostrophic barotropic atmosphere is described by the field $z[x,y]$ of the altitudes of the isobaric surface at a middle level in the troposphere, with z represented as the sum of its Fourier components). Let $d^2[\omega',\omega'']$ be some quantitative measure of the difference between any two states ω' and ω'' (e.g., the square of the difference of the velocity vector fields

$$|\mathbf{u}'(x,y) - \mathbf{u}''(x,y)|^2 = \frac{g^2}{l^2}|\nabla z'(x,y) - \nabla z''(x,y)|^2,$$

averaged over all x and y). The evolution of the state of the atmosphere is described by the dynamic equation $\omega(t) = A^t \omega(0)$, where A^t is a certain operator [for example, for a quasi-geostrophic barotropic atmosphere the dynamic equation has the form of (7.6')]. As the initial data contain random errors, let us assume that there is a distribution of probabilities $P_0(d\omega)$ given at the initial instant $t = 0$ for the set of possible states $\Omega = \{\omega\}$. Then the function $\omega(t)$ will be *random*, and it will be possible to determine the mean square error for the prediction of components of scale L for the period t from the formula

$$\sigma_L^2(t) = \overline{d^2[\omega_L(t),\overline{\omega_L(t)}]}, \tag{20.1}$$

where the bars denote the mathematical expectation. Further, let a *climatic* distribution of probabilities $P(d\omega)$ be assigned for Ω, so that the *climatic* variance of components of scale L is also known, i.e., the quantity

$$\sigma_L^2 = \langle d^2[\omega_L,\langle\omega_L\rangle]\rangle, \tag{20.2}$$

where the angle brackets signify climatic averaging. Then the predictability limit t_L of components of scale L will be the upper limit of those values of t for which the condition

$$\sigma_L^2(t) < \sigma_L^2$$

is satisfied.

The concept of the evolution of the state of the atmosphere as a random process $\omega(t)$ makes it possible to approach the task of studying the possibilities of the statistical extrapolation of this process. Lorenz's papers [138-141] were devoted to this task. A hypothesis expressed by Kolmogorov in a 1967 report at the All-Union Symposium on Turbulent Mechanics in Kiev here acquires theoretical significance. According to this hypothesis, a random process $\omega(t)$ describing the evolution of the turbulent flow in an environment with vanishing viscosity $(v \to 0)$ asymptotically approaches a Markov process for large t (i.e., the distribution of probabilities $P_t(d\omega)$ for $t > t_1$ becomes uniquely determined by the state $\omega(t_1)$ and does not depend on the remote history of the process, i.e., during $t < t_1$). By considering the meteorological fields only on a finite spatial grid of points separated by $L \gg (v^3/\varepsilon)^{1/4}$, we thereby assume negligible viscosity $(v \ll \varepsilon^{1/3} L^{4/3})$ and can assume that the finite-difference random process $\omega(t)$ obtained in this approximation of the meteorological fields will asymptotically approach a Markov process. A Markov scheme was used, for example, by Bugayev, Dzhordzhio, and Sarymsakov[142] for the description and forecast of the time alternation of the types of weather over central Asia and by Chaplygina[143] to describe the alternation of Dzerdzeyevsky's types of circulation.

The predictability problem was first presented in meteorological literature by Thompson,[144] who also made an attempt to calculate the function $\sigma_L^2(t)$ analytically within the limits of the simplest (quasi-geostrophic) prognostic models. A more accurate calculation of this sort was soon performed by Novikov;[145] he used the barotropic model (7.6') (disregarding the term with $1/L_0^2$ and the variability of the Coriolis parameter l) in a case where the main initial field and the initial error field are statistically independent, homogeneous, and isotropic random fields with correlation radii L and L_1; for the average kinetic energy of the error field $E_1(t)$, he obtained the equation

$$\frac{E_1(t)}{E_1(0)} = 1 + 2\frac{(t/T)^2}{[1 + (L/L_1)^2]^2} + \cdots, \tag{20.3}$$

where $T = L/U$ is the typical time scale for the synoptic processes. Setting $T = 1$ day and $L = 2L_1$, we verify that the time t during which the prediction error reaches the random-choice error (so that $E_1(t)/E_1(0) = 16$) turns out to be equal to two weeks.*

The first numerical experiment on predictability was carried out by Diky and Koronatova,[146] who using the barotropic model compared 24- and 48-hour forecasts with real initial data and the same data with the addition of a random error. They took a spatially uncorrelated error field as well as a field with a correlation radius of two grid intervals; they used mean square errors of 1 and 2 dkm. In different series of forecasts the results came out differently, but on the average the error grew by 30–50% in 24 h, with the spatially correlated errors growing somewhat more rapidly than the uncorrelated errors.

Much more complete numerical experiments on predictability were done by Charney[15] with the aid of Smagorinsky's,[66] Leith's,[78] and Mintz's[73] models. The time variations in the prediction errors turned out to be most natural in Mintz's model. Here initial errors were introduced into the temperature field, and the resulting forecast errors in the temperature field were calculated at two levels, 1 (about 400 mb) and 2 (about 800 mb), separately for the Northern (N) and Southern (S) hemispheres. Figure 37 shows the changes in the mean square error of the temperature field forecast for an initial sinusoidal error field $\delta T = \sin 6\lambda \cos 11\varphi$ (with an amplitude of 1°C). From this graph it is clear that, in the process of the velocity field's adapting to a perturbation in the temperature field, the error first decreases, then grows exponentially (doubling every 5 days)

---

*Equations for the complete statistical description of the evolution of the error field have been published by V. I. Tatarsky [Izv. Akad. Nauk SSSR, Fiz. Atmosfery i Okeana **5**, 293 (1969)].

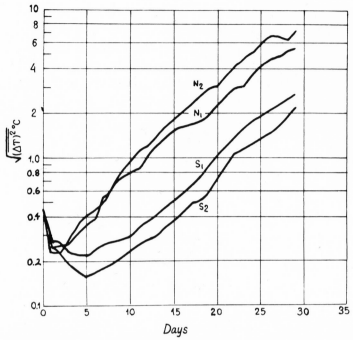

Fig. 37 The mean square temperature forecast error after Charney,[15] using Mintz's model,[73] with a sinusoidal initial error $\delta T = \sin 6\lambda \cos 11\varphi$ (with an amplitude of 1°) that was introduced in Mintz's numerical experiment on the 234th day. N is for the Northern Hemisphere and S is for the Southern; the index 1 is for the level near 400 mb, and 2 is for the level near 800 mb.

according to the linear theory of small-perturbation instability; then, having grown to some finite magnitude, the perturbations enter a regime of nonlinear oscillations. Figure 38 shows the evolution of the forecast error for a random initial error (modulated by the factor $\cos \varphi \cos 6\varphi$), and Fig. 39 shows a case where the initial error is localized in the region 21–63° N, 157–203° W.

For the limit of predictability Charney takes the period in which the forecast error reaches a value equal to the mean square difference between two random temperature fields (or two actual fields: for time differences greater than 3 days their mean square difference turns out to be approximately constant, equal to 5°C in the Northern

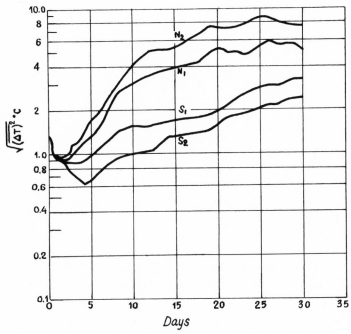

Fig. 38 The mean square temperature forecast error according to Mintz's model[73] with a random initial error (modulated by the factor $\cos \varphi \cos 6\varphi$). After Charney.[15]

Hemisphere and $4°C$ in the Southern Hemisphere at level 1, and $8°C$ and $3°C$, respectively, at level 2; hence it is clear, in particular, that the limiting period of *inertial* prediction of the temperature field is equal to 3 days). If the value of this difference is $8°C$ (at the 800 mb level in the Northern Hemisphere) when the error doubles every 5 days and the initial mean square error is $1°C$, the limit of predictability turns out to be equal to $5 \log_2 8 = 15$ days (Charney explains the identical rate of exponential growth for different kinds of initial errors by the dominating contribution of the baroclinic instability of waves with a wave number along the parallels $m = 6$; these waves are excited for almost any spectrum of initial perturbations because of the strong nonlinear interactions between the different spectral components).

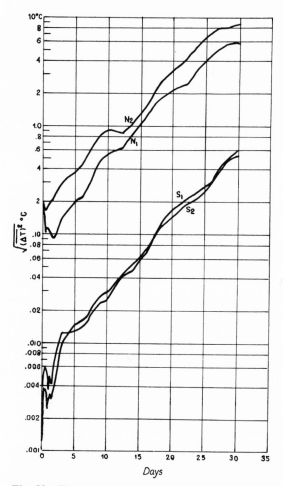

Fig. 39   The mean square temperature forecast error according to Mintz's model with the initial error localized in the region 21–63°N, 157–203°W. After Charney.[15] Note that the gravitational waves that are caused by disturbing the process of velocity and pressure field adaptation in the Northern Hemisphere reach the Southern Hemisphere within 1–2 days; there they create perturbations with an energy three orders of magnitude smaller than that of the original disturbance.

At the International Symposium on a Program of Global Investigation of the Atmosphere (Stockholm, 1967), Smagorinsky reported on the results of a new numerical experiment on predictability. He conducted the experiment with the aid of his nine-level model of the atmosphere. To real initial data he added an uncorrelated field of small, random temperature perturbations on the order of 0.5°C in the average temperature of a column of air. After 14 days, this resulted in only relatively small, small-scale prediction errors (so that the correlation coefficients between the fields predicted from the actual and from the distorted initial data fell, after 14 days, only to 0.99 at a level of 50 mb, to 0.96 at a level of 500 mb, and to 0.88 at a level of 1000 mb). Guided by the character of the growth of the errors at the end of this period, Smagorinsky estimated the predictability limit of the synoptic processes as 3 to 4 weeks, using his model. The increase of this estimate in comparison to Charney's experiments may be related to the considerably increased number of degrees of freedom in the model of the atmosphere employed. If this is so, then since a real atmosphere has many more degrees of freedom than any model, estimates of the predictability limit determined from models should be regarded only as lower limits.*

Determination of the limits of predictability per se is not a constructive task (and should not be an end in itself). A constructive solution of the predictability problem for extended periods must be the determination of the characteristics of the meteorological fields that are predictable for these periods. If individual synoptic processes, cyclones and anticyclones, are not predictable for periods of, say, two, three, or four weeks, then that does not at all mean that it is impossible to predict weather for longer periods: There can exist predictable (for these periods), generalized (averaged) characteristics of ensembles of individual processes that are of great practical

*See also S. Manabe, J. Smagorinsky, J. L. Holloway, Jr., and H. M. Stone, *Monthly Weather Rev.* **98**, 175 (1970).

interest (such, perhaps, are charts of monthly precipitation totals, for example).

A possible way of going beyond the limits of the predictability of individual synoptic processes is the construction of equations for the statistical characteristics of ensembles of such processes, similar to the Reynolds or Friedmann-Keller equations from turbulence theory. The paper by Thompson[147] and my paper[148] can serve as examples; in them were suggested the simplest models for predicting the *zonal* statistical characteristics (the single-point first and second moments) of the meteorological fields. Thus, within the limits of the barotropic model, neglecting the third moments and assuming the approximate isotropy of the nonzonal velocity field, the following equations were constructed in my article[148] for the circulation index $\alpha = \bar{u}_\lambda / a \sin \theta$, the meridional momentum flux (per unit mass) $\tau = \overline{u'_\theta u'_\lambda}$, and the kinetic energy of nonzonal motions (per unit mass)

$$E = \tfrac{1}{2}\left(\overline{u'^2_\theta} + \overline{u'^2_\lambda}\right),$$

where the bar indicates a zonal average, and the prime indicates deviation from the zonal average:

$$\frac{\partial \alpha}{\partial t} = - \frac{1}{a^2 \sin^3 \theta} \frac{\partial \tau \sin^2 \theta}{\partial \theta},$$

$$\frac{\partial \tau}{\partial t} = - E \sin \theta \frac{\partial \alpha}{\partial \theta}, \tag{20.4}$$

$$\frac{\partial E}{\partial t} = - \tau \sin \theta \frac{\partial \alpha}{\partial \theta}.$$

These equations describe a nonlinear propagation (with velocity $E^{1/2}$) of waves along a meridian. Construction of similar equations within the limits of the baroclinic and nonadiabatic models would be of great interest.

Another possible method is the construction of *optimally predictable functionals* of the values of the random process $\omega(t)$ describing the evolution of the states of the atmosphere. Suppose that from the values of the process in the past (for $T_1 \le t \le 0$), we wish to predict its value in the future (for $0 < \tau \le t \le T_2$). Being limited to a *linear* forecast, for some general conditions it is possible to find the *canonical* linear functionals $U_1, U_2, \ldots$ of the values of $\omega(t)$ in the past and $V_1, V_2, \ldots$ of the values of $\omega(t)$ in the future, such that all pairs (U_i, U_j), (V_i, V_j), and (U_i, V_j) will be uncorrelated for $i \ne j$. In this way the correlation coefficients $r_i = r(U_i, V_i)$ will completely describe the statistical dependence of "the future" on "the past."

A method of constructing the canonical functionals and determining the correlation coefficients in the simplest case of a one-dimensional stationary Gaussian random process $\omega(t)$ with a rational spectral density (with $T_1 = T_2 = \infty$) was developed by Yaglom.[149]

Let the numbers $i = 1, 2, \ldots$ be arranged in order of decreasing values of r_i, such that $r_1 \geq r_2 \geq \ldots$. Then V_1 will be the optimally predictable functional. Having assigned the "predictability level" r_0 (e.g., $r_0 = 0.7$), it will be possible to consider all the functionals $V_1, \ldots, V_m$ and $U_1, \ldots, U_m$, for which $r_1 \geq r_2 \geq \ldots \geq r_m \geq r_0$, characteristics of the "weather," and all the remaining functionals $V_{m+1}, V_{m+2}, \ldots$ and $U_{m+1}, U_{m+2}, \ldots$ for which $r_0 \geq r_{m+1} \geq r_{m+2} \ldots$ characteristics of "the turbulence" (Obukhov[135]). It is important that this entire construction, including the treatment of the initial data (i.e., separating from it "the weather" $U_1, \ldots, U_m$ and "the turbulence" $U_{m+1}, U_{m+2}, \ldots$), depends on the forecast period $\tau \leq t \leq T_2$; in particular, the optimum density of the meteorological grid or the optimum scale for smoothing the meteorological fields can depend on it.

A third possible method is to explain only the slow (with typical times on the order of months) processes, of the *joint evolution* of the active layer of the ocean and the atmosphere, which is adapted to the state of the ocean; these create the long-term weather anomalies. Along with this there must be a simplification of the dynamic equations of the A–AL system such that the simplified equations describe the slow processes with sufficient accuracy but their solutions do not include the faster processes of the atmosphere's adaptation to the state of the active layer of the ocean (including the short-term synoptic processes: gyroscopic waves or traveling cyclones and anticyclones). This simplification of the dynamic equations, filtering out the short-term synoptic processes from their solutions, can, roughly speaking, be reduced to neglecting the partial derivatives with respect to time in all equations except those that describe the evolution of the heat content of the active layer of the ocean. Numerical experiments with such simplified equations could help to clarify the feasibility of this approach to long-range weather forecasts.

21. Extraterrestrial Influences

Let us leave aside the modern astrologers who propose to predict from the positions of the planets or the phases of the moon not only the fates of credulous individuals but also the weather (usually for the location at which the astrologer lives; he most often does not consider weather changes at other locations at all). The influence of these factors on the weather is unimportant.

We can mention such influences as *meteor showers*, which perhaps assist in increasing the concentration of ice-producing nuclei in the atmosphere and therefore, eventually, the precipitation. Thus, Maruyama[150] reported observing (from data of 300 weather stations over 60 years) a noticeable increase in the amount of precipitation and especially in the number of rain showers occurring a month after meteor showers; however, at many stations this relationship did not appear (according to the author's explanation, because of dust production during volcanic eruptions or dust transfer from other regions).

But of course, on this point the greatest attention should be devoted to the question of whether there is a connection between the earth's weather and the *fluctuations in solar activity*. The presence of such a connection would be almost a tragedy for meteorology, since it would evidently mean that it would first be necessary to predict the solar activity in order to predict the weather; this would greatly postpone the development of scientific methods of weather prediction. Therefore, arguments concerning the presence of such a connection should be viewed most critically.

Influences of the solar activity on certain *geomagnetic phenomena* (magnetic storms, auroras, etc.) have been established beyond doubt; their most direct explanation is that the plasma clouds radiated by active regions on the sun, brushing against the earth's magnetosphere, create perturbations in it (such plasma clouds present a great threat to cosmonauts). Things seem quite different in regard to the influence of solar activity proclaimed by "helio-geophysics" enthusiastics: Most of the information concerning such an influence (collected in the books by Eygenson,[151, 152] Ruba-shev,[153] and Sazonov,[154] for example) fortunately produces only an impression of successful experiments in autosuggestion; the hypotheses proposed concerning the physical mechanisms of the influence of solar activity on the weather lack convincing sub-stantiation.

No noticeable changes with time are observed in the total energy flux of the solar radiation reaching the earth (hence the term "solar constant"). The solar activity is manifested in the strong variations of the corpuscular component of the sun's radiation, but the oscilla-tions (related to these variations) of the energy influx to the earth are apparently very small in comparison with the energy of cyclones, for example. Since it appears that there is no direct energy influence of solar activity on the weather, those who propose hypotheses for physical mechanisms of the influence of solar activity are reduced to a search for some kind of triggering mechanism in the earth's atmosphere, similar to a switchman who changes the direction of motion of a heavy railroad train with the slight force of his hand; because of such mechanisms, even energetically insignificant variations in the influx of solar radiation could lead to considerable changes in the weather.

The same hope that the atmosphere, like a ropewalker, balances on the edge of an unstable equilibrium (i.e., is near a potential energy maximum, or on a "potential ridge"), and a small push from outside can cause *directed* changes in its state (force it to slide into a "potential well") is cherished by enthusiasts of artificial weather manipulation. However, there are apparently a sufficient number of internal mechanisms in the atmosphere that permit instability but do not allow it to climb too far up the sides of "potential ridges" (unstable formations, e.g., certain types of supercooled or hail clouds, are found relatively rarely and only on smaller scales); the atmosphere is probably most often situated in "potential wells." In such states, only energetically powerful influences can have a great effect, and even then only when natural instabilities are prevented.

The influence of solar activity on the weather would logically be most easily represented as a statistical result of an accumulation of some small effects over long periods of time. The presence of such long-term influences is most conveniently checked in the instance of the 11.5-year cycle of solar activity. This cycle was determined by Wolf[155] in studying oscillations of the index $W = n + 10\,N$, where n is the number of sunspots and N is the number of groups of spots on the visible side of the sun during the period beginning January 1749 (periods of maxima and minima of W had been observed as much as a century earlier—even before 1610). The nature of the oscillations in the average monthly Wolf numbers W is quite clear from the lower curve in Fig. 40, which shows data for the years 1820–1950. At the minima the values of W are only a few units, and at the maxima they reach 80–140 units. The changes

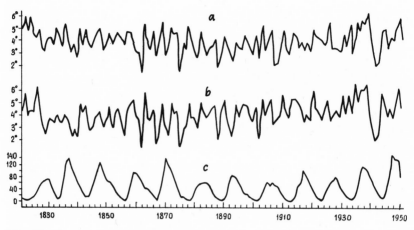

Fig. 40 The average monthly values of the air temperature in Moscow (a) and Leningrad (b) and of the Wolf numbers (c).

of W within the cycles are not sinusoidal: The rising occurs faster (in an average of 4 years) than the sinking (7 years), so that in the oscillation spectrum of W (Fig. 41), along with a maximum for an 11.5-year period there also appear maxima corresponding to harmonics of that period (they correspond to a semiperiod of about 5 years in Fig. 41).

Wolf numbers are apparently not the best index of solar activity for heliogeophysics: They probably characterize to some degree the oscillations of the ultraviolet, but not the corpuscular radiation of the sun. The latter is characterized better, but still not very well, by the total area of the sunspots Σ (the correlation coefficient between Σ and W is 0.85). Many other indices of solar activity that change differently with time are being studied. In order to avoid the difficulties in choosing suitable indices, as well as the subjectivity involved in estimating the similarity between the fluctuations in these indices and fluctuations in weather characteristics (this subjectivity is typical of most of the works referred to[151-154]), it is better to consider only the spectra of weather characteristics and to

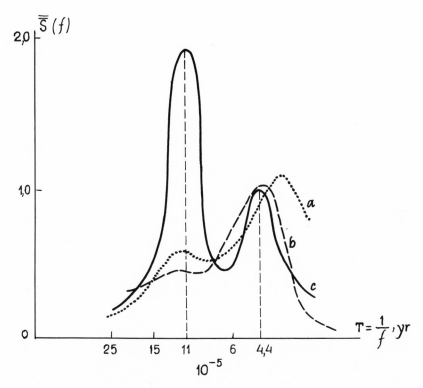

Fig. 41 The spectral densities of the air temperature fluctuations in Moscow (a) and Leningrad (b) and of the Wolf numbers (c). After materials by Kolesnikova and Monin.

ascertain whether periods peculiar to the fluctuations of solar activity are manifested in them.

Short-period oscillations of the various meteorological elements can vary in character; they can essentially depend on local conditions and can therefore be unrepresentative of the world's weather. However, one can hope that in long-period oscillations (interannual, intrasecular, etc.) the difference in the behavior of the various meteorological elements and the dependence on local conditions must weaken, and the spectra of most meteorological elements and planetary indices must become at least approximately similar to each other (Kolesnikova and Monin[134]). For the purpose of checking this assumption as well as the presence of influences by solar periodicity, Kolesnikova and the author examined, to begin with, the 130-year series of air-temperature observations in Moscow and the 198-year series for Leningrad.

Graphs of the oscillations in the average monthly air temperature in Moscow and Leningrad, given in Fig. 40, show that the most pronounced temperature oscillations have significantly shorter periods than oscillations of the Wolf numbers and that there is no close connection between the phases of the oscillations of the temperature and the Wolf numbers. Spectra of the temperature oscillations (Fig. 41) fully support this impression: They *do not show* an 11.5-year period, but there is a weak maximum in the interval of periods between 2 and 5 years. The spectra constructed by Landsberg, Mitchell, and Crutcher[156] of temperature oscillations at several stations in Maryland and the spectra of 100- to 160-year series of annual precipitation totals at 8 European stations according to Brier[157] have a similar appearance; the spectrum of the series obtained by summing over the 8 stations *does not* show any periodicity for 22–23 years, 11–12 years, or 5–6 years.

Thus, the fundamental 11.5-year period of the solar cycle is not manifested in spectra of some meteorological elements. The data available on fluctuations of the solar activity are quite insufficient for studying still longer periods. So, for example, the assertion by enthusiasts of "heliogeophysics" that an 80- to 90-year cycle of solar activity exists, formulated on the basis of graphs like that of Fig. 42 and containing *only two* such cycles, is of course groundless: The well-known tendency of the human psyche to regard three to five repetitions of some cycle as a revelation of some periodicity is not corroborated by mathematical statistics, and here the cycle is repeated only twice.

It would also be groundless to be troubled by the coincidence of the growth of Wolf numbers during the first half of the twentieth century (again, see Fig. 42) with the warming that took place: The

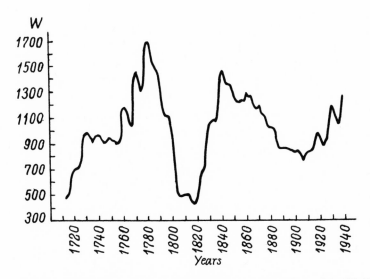

Fig. 42   Oscillations of the 23-year floating average values of the Wolf numbers. After Rubashev.[153]

same growth of Wolf numbers a hundred and two hundred years ago was apparently not accompanied by similar warming, and there is no obvious similarity at all between Fig. 42 and Fig. 32, which describes the intersecular fluctuations of the climate.

Note, finally, that meteorologists are by no means forced to make hypotheses concerning the influence of solar activity on the weather. Obviously, the weather changes on earth, including long-term changes, will take place with the same amplitudes as those actually observed even when solar activity is completely absent, i.e., when the entire spectrum of solar radiation is strictly constant.

Numerical experiments on the general circulation of the atmosphere confirm this, at least for intra-annual weather oscillations, and the discovery of long-term processes similar to the ocean-atmosphere interactions studied by Bjerknes makes it possible to dispense with hypotheses concerning solar influences in the inter-annual oscillation interval as well.

4

Additions

22. Modeling the Planetary Circulation

In modeling the general circulation of the atmosphere or the ocean in a laboratory setting, difficulties arise even in the very first stage— in attempts to guarantee the *geometric similarity* of the model and the actual phenomenon. In fact, the general circulation has scales that are tens of thousands of kilometers ($L \sim 10^7$ m), so that for a model with dimensions on the order of meters, the scale factor is equal to 10^{-7}; this means that the layer of fluid in the model that corresponds to the troposphere (with thickness $H \sim 10$ km) would be only 1 mm thick, which is quite inconvenient for experimentation. In practice the atmosphere is modeled by a layer of fluid with a thickness of about 10 cm, so that the number H/L turns out to be two orders of magnitude greater in models than in reality; it is clear that in such models it is possible to study only those phenomena that have little dependence on the value of the parameter H/L when its value is small.

It is obvious that with a scale factor of 10^{-7}, motions with scales smaller than 10 km (including all small-scale turbulence) are not modeled. The impossibility of ensuring the geometric similarity of motions on the entire scale spectrum (from $L \sim 10^7$ m to the internal turbulence scale $\lambda \sim 1$ mm) results in the loss of dynamic similarity as well. Thus, the Reynolds number $\mathrm{Re} = UL/v \sim (L/\lambda)^{4/3}$ for the general atmospheric circulation has a magnitude of the order of 10^{13}; but in laboratory models of the general circulation, it is convenient to have subcritical values of Re on the order of 10^2 to 10^3, in order to prevent the turbulence from obscuring the stream-lines of the general circulation.

The discrepancy between the values of Re in nature and in models can be alleviated by assuming that the small-scale turbulence in nature is analogous to the molecular viscosity in models in its influence on large-scale flows. The value of Re for the real circulation can then be calculated by using the small-scale turbulent viscosity $v_{\mathrm{turb}} \sim 10^4$ to 10^5 cm^2/sec. The number Re will then decrease by five to six orders of magnitude, but for the real circulation Re will still remain much higher than in models. In such models it is possible to study only those phenomena that have little dependence on the value of Re when its value is large, i.e., those that have *similarity with respect to the Reynolds number*.

This same reasoning applies to the dimensionless parameters related to Re that describe the large-scale thermal convection in the atmosphere: the Péclet number $Pe = UL/\chi = Re \cdot Pr$ (χ is the coefficient of thermal conductivity, $Pr = \nu/\chi$ is the Prandtl number), the Nusselt number $Nu = qL/\kappa\delta T$ (q is the heat flux caused by convection and δT is the temperature drop in the convection layer), the Grashof number $Gr = (gH^3/\nu^2)(\delta T/T)$ (where H is interpreted as the thickness of the convection layer), and the Rayleigh number $Ra = Gr \cdot Pr$. The relationship of Nu and Re is obvious from the fact that $Nu = L/H \sim Re^{1/2}$ is obtained when $q = \kappa\delta T/H$ (the equation $H \sim L/Re^{1/2}$ is the usual estimate of boundary-layer theory). The relationship of Gr to Re consists in the relation $Gr = Re^2$, which is obtained by using a scale of length H in Re and defining the typical speed of convective motions by the equation $U = (gH\delta T/T)^{1/2}$. Even in calculating the values of all these parameters for the actual circulation by using the coefficients of turbulent viscosity and thermal conductivity ν_{turb} and χ_{turb}, these values turn out much larger than in the models. Therefore, one can only hope that the phenomena of large-scale thermal convection have little dependence on the values of the indicated parameters when those values are large.

The most important parameter determining the character of large-scale atmospheric motions is the Kibel number $Ki = U/Ll$ (here, in the case of rotating models, it is necessary to put $l = 2\omega$, where ω is the angular velocity of rotation). It is sometimes convenient (see Kuo[1]) to take as U the typical speed $(R/l)(\delta_h T/L)$ of the so-called thermal wind, which is created by the horizontal temperature differences $\delta_h T$, so that for $H = RT/g$ we obtain $Ki_T = (gH/L^2 l^2)(\delta_h T/T)$. For large-scale atmospheric motions this parameter has a magnitude on the order of 10^{-1}, and for the ocean it is apparently on the order of 10^{-3}. Such values are not hard to obtain in rotating models as well (by choosing ω as the speed of rotation). This is what guarantees the success of laboratory modeling of the planetary circulation.

Finally, the hydrostatic stability of the atmosphere or ocean can be described, for example, by the Richardson number $Ri = (gH/U^2)(\delta T/T)$ (the ratio of the energy of convective motions to the energy of the wind). For the general atmospheric circulation it has values on the order of 10^2, also achieved in laboratory models; in the ocean, on the other hand, it is considerably larger (by a factor of 10^3). For a more detailed discussion of similarity criteria for laboratory modeling of the planetary circulation, see the review works by Fultz,[2-4] Fultz et al.,[5] Arx,[6] Dmitriyev, Bonchkovskaya, and Byzova,[7] and Bonchkovskaya.[8] Certain phenomena and peculiarities of the general circulation of the atmosphere and the ocean are successfully reproduced in laboratory models. Some of them are listed below.

1. Taylor and Proudman's early experiments (1917–1923) with rotating vessels (see the review by Squire[9]) demonstrated the great *stability of geostrophic vortices*, which behave in many ways like elastic bodies: In particular,[5] this stability suppresses a tendency for excessive vertical currents to occur in the model as a result of the overestimation of the ratio H/L in it; i.e., it weakens the dependence of the motions on this parameter.

2. By placing small obstacles into a rotating hemisphere filled with fluid, Long, Fultz, and Frenzen (1951–1955) succeeded in generating *gyroscopic Rossby waves*.

3. Arx (1952–1957) modeled the *western intensification of ocean currents* (resulting from the β-effect) in a rotating layer of fluid that was parabolic or flat but in which the depth increased with the radius.

4. Vettin,[10] in 1857–1885, reproduced Hadley's *trade circulation* in a rotating cylinder with heating on the outside and cooling along the axis.

5. Fultz (1951) established that, in a rotating cylinder with heating on the outside and cooling along the axis, an axially symmetric (trade) circulation occurs only when $\mathrm{Ki} > \frac{1}{2}$; but for small Ki it loses its stability, and irregular, nonstationary currents are produced, often in the form of *narrow streams and geostrophic vortices*, similar to the large-scale atmospheric motions of the middle latitudes (Fig. 43). The dimensionless profiles of the zonal velocity and of

Fig. 43 Flow lines of the surface currents of water in a rotating cylindrical vessel with heating along the edges. After Fultz et al.[5] The external radius $L = 15.7$ cm; the depth of the layer of water $H = 6$ cm; the speed of rotation $\omega = 0.75$ sec^{-1} in a counterclockwise direction; the heating was begun 3.5 min before measurement; the temperature of the water is $33°$ C; in the streams $Ki_T = 0.15$ and overall $Ki_T = 0.02$; Ri ~ 50; Re ~ 1400; regions of fastest jet streams are shaded.

the turbulent momentum flux turn out to be close to that observed in the atmosphere. Similar results have been obtained with models of the same kind by Bonchkovskaya;[7] later she also constructed models with an inhomogeneous (zonal and nonzonal) warming of the bottom; moreover, on a model with zonal heating, she succeeded in modeling fairly well the zonal profiles of the temperature and the coefficient of macroturbulent heat conductivity.

6. Fultz experimented (1952–1956) with a rotating cylindrical vessel with two layers composed of fluids with different densities. For large Ki, he succeeded in modeling a sloped (axially symmetric) interface with a slope described by Margules's equation. For small Ki he modeled the *instability of a frontal surface*: Regular waves formed

on it, turning into cyclones with cold fronts and then an occlusion; later, irregular currents like those of Fig. 43 developed.

7. In a model proposed by Hide[11] in 1950—a current between two coaxial cylinders rotating with the same angular velocity, with one cylinder heated and the other cooled—it is possible to successfully generate *standing gyroscopic waves* with any given (of course, not too large) wave number m. Gradually decreasing Ki_T (by decreasing the radial temperature difference $\delta_h T$) with a fixed rate of rotation (the Froude number $Fr = L\omega^2/g$), one can observe discrete transitions $m \to m + 1$ [see the curves separating the regions with different wave patterns in the plane (Fr, Ki_T) in Fig. 44]. It is interesting to note that these transitions exhibit a strong hysteresis: The transition $m + 1 \to m$ takes place with a considerably larger Ki_T than does $m \to m + 1$. A theoretical calculation of these curves was carried out in Kuo's 1957 paper.[1] A calculation of the velocities in the field of standing waves (carried out by Riehl and Fultz[12] for $m = 3$) demonstrated that the mean meridional circulation here consists of three cells; the dimensionless vertical velocities and moreover the profiles of the meridional heat fluxes produced by the mean circulation and the macroturbulence turn out to be close to those observed in nature.

8. In the model mentioned above, Hide[11] found quasi-periodic oscillations in the intensity and form of waves and other flow characteristics (in limiting cases, these are the transitions $m_1 \rightleftarrows m_2$) with periods on the order of 10–200 cycles of the cylinders; these oscillations apparently model the *index cycle*. Oscillations with much longer periods are also observed (see the example in Fig. 45), arising without any noticeable periodic external influences; this should be of special interest for the interpretation of long-period oscillations of the general circulation of the atmosphere and ocean.

Reproduction of the basic features of some known phenomenon in models still does not in itself add anything to our knowledge. However, creating opportunities that nature does not offer (e.g.,

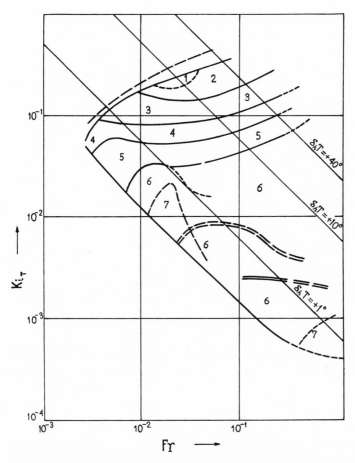

Fig. 44 Regions with different wave patterns characterized by wave numbers m in the flow between coaxial cylinders (with radii $L/2 = 2.46$ and $L = 4.92\,\text{cm}$ and height $H = 13\,\text{cm}$). The cylinders rotate with the same angular velocity ω (the abscissa is the Froude number $\text{Fr} = L\omega^2/g$), while there is heating of the outer and cooling of the inner cylinder (the ordinate is the thermal Kibel number $\text{Ki}_T = (gH/L^2\omega^2)\,(\delta_h T/T)$, with $T \approx 21^\circ\text{C}$ and $\delta_h T$ gradually increasing). After Fultz et al.[5]

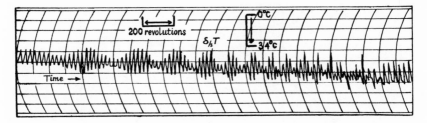

Fig. 45 Oscillations of the radial temperature difference during a very slow increase of that difference (Ki_T changes from 0.07 to 0.09) in the flow between coaxial cylinders rotating with the same angular velocity ($\omega = 3\,\mathrm{rad/sec}$). After Fultz.[4] The high-frequency oscillations are the index cycle, with a period of about 30 "days" (rotations of the cylinders); their envelope at first has a period of about 400 "days," and then it shortens.

obtaining the standing gyroscopic waves with a fixed wave number) or discovering phenomena whose analogs had not earlier been noticed in nature (e.g., the long-period oscillations modulating the "index cycle") makes it possible to find out something new. From the foregoing it is clear that from this point of view rotating models with heating open up interesting perspectives for studying certain features of the general atmospheric circulation.

23. The Atmospheres of Other Planets

The most important parameters determining the character of a planet's atmosphere are the following: (1) *the thickness of the atmosphere*, described by the altitude of the homogeneous atmosphere $H = p_0/\rho_0 g$ [which is related to the sound speed $c_0 = (\kappa g H)^{1/2} = (\kappa R_0 T/\mu)^{1/2}$ and the typical vertical temperature gradient $\delta T/H$, which should be comparable to the adiabatic lapse rate $\gamma_a = [(\kappa - 1)/\kappa]\,(g\mu/R_0)$, where μ is the molecular weight and R_0 is the universal gas constant, in order to judge the hydrostatic stability of the atmosphere]; (2) *the rate of the planet's rotation* ω (and the related coefficient of horizontal compressibility $\omega a/c_0$, which is proportional to the ratio Ma/Ki); (3) *the transparency of the atmosphere* (determined by the composition of the atmospheric gases and the aerosol) with respect to radiation of various wavelengths.*

The latest information on the planets of the solar system, information necessary for estimating the parameters indicated above, can be found in the book by Kellogg and Sagan[13] (which contains, in particular, Mintz's work on the theory of the general circulation of planetary atmospheres), the review by the same two,[14] the collection edited by Kuiper and Middlehurst,[15] and the quite recent book by Moroz.[16] However, more accurate information is presently being obtained at a rapidly accelerating rate. A lot of very important new data were obtained by the splendid experiment of the Soviet unmanned probe *Venus 4*, which on 18 October 1967 made a smooth descent into the atmosphere of Venus and transmitted measurements of the vertical profiles of Venus's atmospheric characteristics to Earth. Finally, two works on the atmosphere of the sun should be mentioned: a review by Jager[17] and an article by Ward[18] on the circulation of the solar atmosphere.

Basic data on the parameters of planetary atmospheres are given in Tables 1–3, which are on the whole adopted from a report by Golitsyn and Moroz at the 1964 Conference on the General Atmospheric Circulation in Moscow. Table 1 gives the radii of the planets a (in km), the periods τ of their revolution around the sun (in days), the periods $2\pi/\omega$ of their own rotation (in days), the angles φ_0 of the obliquity of their ecliptics, the acceleration of the force of gravity g at the surface of each planet (in cm/sec^2), and the albedo A of each planet. It is important to keep in mind that the *solar day* on Mercury is about 180 days (one sunrise in two of Mercury's years), and on Venus with its backward rotation it is 120 days (two sunrises in one of Venus's years). Jupiter is noticeably flattened: The equatorial radius is appreciably larger than the polar radius, and the force of gravity is appreciably smaller at the equator than at either

---

* Methods of estimating the circulation in the atmospheres of planets from external parameters have been suggested by G. S. Golitsyn [*Izv. Akad. Nauk SSSR, Fiz. Atmosfery i Okeana* **4**, 1131 (1968); *Dokl. Akad. Nauk SSSR* **190**, 323–326 (1970)].

Table 1 Astronomical Characteristics of the Planets

Planet	a (km)	τ (days)	$\frac{2\pi}{\omega}$ (days)	φ_0	$g\left(\frac{cm}{sec^2}\right)$	A
Mercury	2434	88	59	$<28°$	388	0.09
Venus	6053 ± 3	225	-243	$1.2°$	884	0.77 ± 0.07
Earth	6371	365	1	$23°27'$	981	0.38
Mars	3370	687	1.02	$24°57'$	370	0.22
Jupiter	71,600–67,900	4339	0.41	$3°7'$	2360–2600	0.48
Sun	690,000	—	25–30	$7°15'$	27,400	—

Table 2 Quantity and Composition of Planetary Atmospheres

Planet	p_0 (atm)	$\frac{p_{CO_2}}{p_0}$	$\frac{p_{H_2O}}{p_0}$	μ	$\gamma_a\left(\frac{°C}{km}\right)$
Mercury	≤0.001	0.1–1.0	—	30–44	3–4.5
Venus	20–100	0.8–1.0	1×10^{-3}–7×10^{-3}	44	8.5–11
Earth	1	3×10^{-4}	2×10^{-3}	29	10
Mars	0.01	0.8–1.0	$(5\pm3)\times10^{-5}$	41–44	4.5
Jupiter	1.3	—	10^{-3}–10^{-2}	2.6	2.5
Sun	0.05	—	—	1	13.4

pole. The period of rotation of its atmosphere is slightly shorter (by 5 min) at the equator than in the middle latitudes; the same effect is much more sharply pronounced on the sun: The frequency of rotation ω (in degrees of longitude per day) varies with the latitude φ according to the formula $\omega = 14.38° - 2.77°\sin^2\varphi$.

Table 2 gives data on the quantity and composition of the atmospheres on the planets: the total atmospheric pressure p_0 in the atmospheres (at the upper cloud surface for Jupiter), the CO_2 and H_2O contents (CH_4 and NH_3 for Jupiter), the molecular weight μ, and the adiabatic temperature gradient γ_a in °C/km. The data for Venus are based on *Venus 4* data; in calculating γ_a, it has been assumed that $\kappa = c_p/c_v = 9/7$ in the upper atmosphere of Venus and $\kappa \approx 1.22$ in the lower atmosphere. The data for Mars take those from the American *Mariner 4* into account. The lower boundary of the sun's atmosphere is chosen arbitrarily.

Table 3 gives data on the solar constant S_0 in cal/cm²-min, the temperature T_1 of the illuminated side of the planet in °K (for Jupiter T_1 is the brightness temperature of the atmosphere in a band from 8–13 μ), the temperature T_2 of the dark side of the planet (for Mercury, at the point opposite the sun, and for Venus, Earth,

Table 3 Thermal Characteristics of Planetary Atmospheres

Planet	S_0 $\left(\dfrac{\text{cal}}{\text{cm}^2\text{-min}}\right)$	T_1 (°K)	T_2 (°K)	c_0 $\left(\dfrac{\text{m}}{\text{sec}}\right)$	H (km)	$\dfrac{a\omega}{c_0}$
Mercury	13.4	620	100–150	440–540	30–45	6×10^{-3}
Venus	3.8	650–750	550	370–410	6–12	6×10^{-3}
Earth	2	280	240	340	8	1.4
Mars	0.86	280	200	270	14	0.83
Jupiter	0.074	130	130	790	17	17
Sun	—	5800	—	9200	160	0.21

and Mars, at the poles), the sound speed $c_0 = [\kappa (R_0/\mu) T_1]^{1/2}$ in m/sec, the thickness of the homogeneous atmosphere at the surface of the planet $H = R_0 T_1/\mu g$ in km, and the dimensionless parameter $a\omega/c_0$. Many of the figures in the tables will require future correction, but the order of magnitude of the parameter $a\omega/c_0$ can apparently be trusted.

Evidently the parameter $a\omega/c_0$ is most important for the character of the atmospheric circulation. In cases where it is small (Mercury and Venus) the rotation of the planet has almost no effect on atmospheric motions; the circulation in such cases is perhaps determined, not by the temperature difference between the equator and the poles, but by the difference between the points closest to and farthest away from the sun. For intermediate values of $a\omega/c_0$ (Earth, Mars, the sun) the character of the atmospheric circulation can be determined by the rotation as well as by the other factors (e.g., the difference in the properties of the oceans and the continents on Earth). For large $a\omega/c_0$ (Jupiter, Saturn, Uranus, and Neptune) the planet's rotation must play the decisive role.

In particular, the behavior of gyroscopic waves, which are the most important of the synoptic processes determining the weather on Earth, depend strongly on the value of $a\omega/c_0$ (Diky,[19, 20] Golitsyn and Diky[21-23]). If for small $a\omega/c_0$ the periods of the gyroscopic waves are determined by the well-known formula

$$\tau = \frac{\pi n(n + 1)}{m\omega}$$

[obtained from formula (5.6) for $\alpha = 0$; here m and n are the longitudinal and latitudinal wave numbers), then for large $a\omega/c_0$ the periods τ will be determined from the relation

$$2\frac{\omega a}{c_0}\left(\frac{\omega\tau}{\pi}\right)^{-2} + \frac{m}{2}\left(\frac{\omega a}{c_0}\right)^{-1}\frac{\omega\tau}{\pi} = 2N + 1, \tag{23.1}$$

which is fully analogous to Bohr's quantum equation (N is any positive integer). As in the solution of the quantum-mechanical problem of determining the eigenvalues of a wave function in the quasi-classical approximation, for large $a\omega/c_0$ the asymptotes for τ turn out to be of the form

$$\tau_n^n = \pi\sqrt{\frac{2a}{\omega c_0}}\left(1 + \frac{n}{2}\sqrt{\frac{c_0}{2a\omega}}\right),$$

$$\tau_n^m = 4\pi\frac{a}{c_0}\left(\frac{n + \frac{1}{2}}{m} - 1\right).$$

$$(23.2)$$

For example, for Jupiter we obtain $\tau_1^{\,1} = 3.2$, $\tau_2^{\,1} = 52$, $\tau_2^{\,2} = 3.5$, $\tau_3^{\,1} = 86$, $\tau_3^{\,2} = 27$, and $\tau_3^{\,3} = 3.8$ of Jupiter's days (in two articles by Golitsyn and Diky [21,22] there are also calculations of the functions of latitude that describe the form of the corresponding gyroscopic waves).

Some planets should be discussed in more detail. With more certainty in estimating the characteristic rotation of Venus and after the appearance of *Venus 4* data testifying in favor of the greenhouse model of its thick atmosphere (consisting of CO_2 with water clouds of convective origin), it is now becoming possible to carry out numerical experiments on its atmospheric circulation, which is evidently quite unlike that of Earth (in particular, it is essentially nongeostrophic, which, however, does not preclude the quasi-solenoidal character of large-scale motions).

Some observational data on the atmospheric circulation on Mars should be mentioned: (1) the numerous data on the seasonal variations of the white polar caps; (2) the charts of isotherms of the surface of Mars, constructed by Hess[24] and Gifford[25] from data of radiometric measurements in the "transparency window" from $8-13\,\mu$ (see the example in Fig. 46); (3) the infrequent observations of movement of the thin clouds (high blue and white ones, which may be ice, and the low yellow ones, which are probably dust).

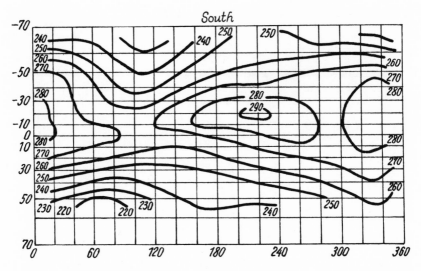

Fig. 46   The average temperature distribution of the Mars surface during winter in the southern hemisphere. After Gifford.[25]

From 18 observations of cloud movements and from isotherms like those of Fig. 46, Hess[24] even constructed hypothetical charts of streamlines; in addition, Gifford[26] analyzed another 36 observations of cloud movement, practically exhausting all the available data of this type.

Several authors have designed models of the vertical structure of the Martian atmosphere. For example, see the article by Prabhakara and Hogan[27] and the three models proposed by Moroz:[16] the maximal ($p_0 = 20$ mb; $0.1 \, CO_2 + 0.9 \, N_2$), the mean ($p_0 = 10$ mb; $0.5 \, CO_2 + 0.25 \, N_2 + 0.25 \, Ar$), and the minimal ($p_0 = 5$ mb; 100% CO_2); the temperature and density profiles according to these models are pictured in Fig. 47. Finally, there is the paper by Leovy and Mintz,[28] who, using the minimal model indicated above, conducted a numerical experiment on the general atmospheric circulation on Mars for winter in the northern hemisphere. They used Mintz's[29-32] two-level model (described in section 18) with the primitive equations. A map of the albedo of the surface of Mars

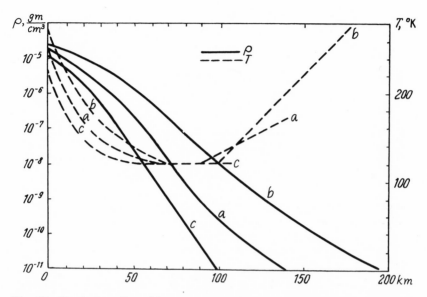

Fig. 47 Vertical profiles of the temperature T and the density ρ in the Martian atmosphere according to the average (a), maximum (b), and minimum (c) models of Moroz.[16]

was assigned. At the initial instant the atmosphere was assumed to be at rest with a constant temperature of 200°K, and the polar caps were absent. The experiment was computed for 24 days.

In this experiment the thin atmosphere of Mars gathered momentum in 7 days, reaching a state of statistical equilibrium with wind velocities of about 25 m/sec in which appeared a cycle of oscillations with a period of 6 days (possibly an analog of the "index cycle"). In the northern (winter) hemisphere a westerly jet stream formed and a *wave pattern* developed with the predominant wave number $m = 3$ (most noticeable on charts of the upper-level temperature; see Fig. 48). In the southern hemisphere weaker easterly currents developed, and in the tropics there was a cell with direct meridional circulation. The southern hemisphere clearly showed the effect of a *daily tide*. On the surface of Mars there formed an average zonal temperature profile that decreased monotonically from 250°K at the south pole

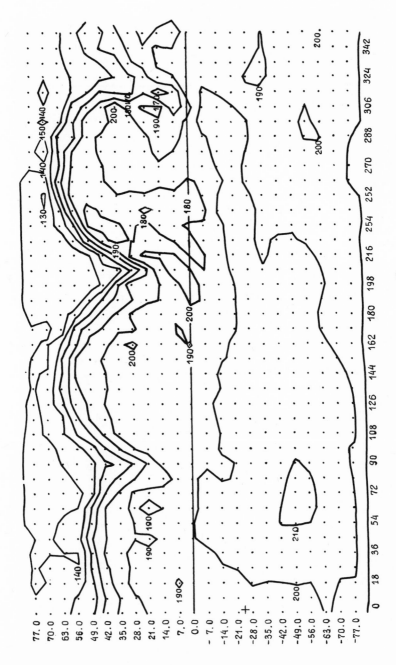

Fig. 48 The temperature field at an upper level in the Martian atmosphere (the altitude is about 12 km) for the 19th day of the numerical experiment by Leovy and Mintz.[28]

to 143.6°K (the temperature of condensation of CO_2) north of 60° N (the polar cap of frozen carbon dioxide). In spite of its preliminary nature, this first numerical experiment on the general atmospheric circulation of other planets has undoubtedly yielded some very intriguing results.

The atmospheric circulation can be most clearly observed on Jupiter (Fig. 49), where dark, reddish bands and the Great Red Spot are clearly visible. The bands are parallel to the equator and separate bright areas that are apparently cloud zones (perhaps formed of condensed NH_3 in a hydrogen-helium atmosphere). The spot is in the southern tropical zone and has the dimensions 1.3×10^4 km by 4.0×10^4 km. Usually five zones are visible, but

Fig. 49 A photograph of Jupiter in blue light at 0905 UT, 23 October 1964, obtained by R. B. Minton of New Mexico State University.

their number, configuration, and details change with time (sometimes at tremendous rates, when the shape of details with dimensions on the order of 10^4 km changes within 10^3 sec; these are probably phenomena of the colloidal instability of the clouds). These changes show a periodicity on the order of 12–16 and 3–6 years, apparently unrelated to the cycles of solar activity.

Perhaps the most peculiar feature of the zonal atmospheric circulation of Jupiter is the presence of abrupt changes along a meridian in the speed of rotation of the atmosphere; there are about 20 of these jumps, which are as large as 1 km/sec in 3° of latitude (at the border of the equatorial zone). The total thickness of Jupiter's atmosphere is very great, apparently no less than 1000 km (on the other hand, according to Peebles's[33] model, hydrogen on Jupiter goes into a metallic state at a depth of $0.2a \approx 14{,}000$ km), and solar radiation penetrates only to a depth on the order of 20 km, so that the atmospheric circulation is evidently created, not by solar heating, but by internal (gravitational) sources of energy. Finally, Jupiter is a powerful source of radio waves, which helped to establish that there are radiation belts and a magnetic field with a field intensity of about 50 G.

Papers by Hide,[34-38] Link,[39] Stone,[40] and other authors have been devoted to the analysis of Jupiter's atmospheric circulation and to attempts at explaining it. Hide proposed an interpretation of the Great Red Spot as a disturbance in the geostrophic flux created by an irregularity at the hard surface of the planet. If the altitude of the irregularity h exceeds $H_0 \cdot \mathrm{Ki}$ where H_0 is the total thickness of the atmosphere, then such a disturbance will form a very stable "Taylor column" piercing the entire atmosphere. If on Jupiter $H_0 \sim 1000$ km, and $\mathrm{Ki} \sim 10^{-3}$, then in order to produce a "Taylor column" it is sufficient to have an irregularity with an altitude $h > 1$ km. The variations of the period $2\pi/\omega$ observed in the zonal rotation of the Great Red Spot ($9^h\,55^m\,38^s \pm 7^s$) could be explained by the transfer of momentum between the hard planet and its thick atmosphere. The dense lower layers of the atmosphere can have such a large electric conductivity that their interaction with the hard planet can be determined not so much by the usual viscosity as by the magnetohydrodynamic viscosity.

To conclude the present section, let us discuss the circulation in the sun's atmosphere. The equatorial current on the sun has a velocity of 2 km/sec. Ward[18] used sunspots as indicators of movements in the solar atmosphere and worked with the Greenwich data on sunspot movements for 1935–1944. Ward states that the motions with scales that are small in comparison to the sun's radius but large in comparison to the dimensions of sunspot groups have meridional and zonal velocities on the order of tens of m/sec with a structure similar to that of gyroscopic waves and that regular meridional circulations similar to Hadley's cells are not observed. These motions transfer the angular momentum from the middle latitudes to the equatorial zone, i.e., against the angular velocity gradient; the kinetic energy of these motions is then transformed into the energy of the equatorial current.

The typical picture of motions within individual sunspots is one of cyclonic rotation (convergence deflected by the Coriolis force) at levels more than 500 km above the photosphere and anticyclonic rotation (due to divergence) in the photosphere with velocities on the order of several kilometers per second. In granules (sizes on the order of hundreds of kilometers, lifetime on the order of minutes) the velocities of convective movements vary from fractions of a kilometer per second in the photosphere to several kilometers per second in the chromosphere and 10–20 km/sec at altitudes of thousands of kilometers above the visible surface of the sun (the photosphere).

24. Planning and Forecasts

Any planning is to a large extent based (or at least should be based) on forecasts. In a discussion of such future factors as may be controlled by actively influencing them, planning and predicting simply coincide. However, when the discussion turns toward consideration of factors of the future that cannot be controlled by us, e.g., elemental weather phenomena, then the connection between the available forecasts and the plans that are being worked out often turns out weak and is sometimes completely absent, leading to a considerable decrease in the economic efficiency of the plans that have been made.

To be specific, let us consider the hydrometeorological phenomena: the weather, the formation and breakup of ice in rivers, river level fluctuations, flash floods, and so on (but it should be stressed that most of our reasoning can relate not only to hydrometeorological forecasts but to many other kinds as well). Agriculture, aviation, (oceanic, river, and even rail and automobile) transportation, hydroenergetics, construction, many branches of mass public service, and other forms of human activity are subject to considerable influence by the elemental phenomena of nature, so that predictions of such phenomena must be considered in planning work in these branches of the economy.

However, many of those who use hydrometeorological forecasts frequently limit themselves to only a passive acquaintance with forecasts and do not give them their due consideration in planning activities. Attempts are sometimes made to justify such a wasteful, narrow-minded attitude toward forecasts by referring to their unreliability without understanding that economic benefits can be gained even from unreliable forecasts (if, of course, the methods of prediction take into account the actual relations among natural phenomena, at least to some extent).

Increasing the reliability of forecasts is undoubtedly the foremost problem of hydrometeorological science, but a no less important task of the hydrometeorological services must be the organization of the rational, economic use of today's forecasts by their various users. Service to specific users must not be limited to attempts at some elaboration of the forecasts, as it is today, but must include acquainting the users with characteristics of the degree of reliability of forecasts, aiding them in estimating the economic dependence of one or another sector of the economy on the phenomena predicted, and recommending the optimum strategy for using forecasts in planning activities.

Following the presentation in two of my papers[41,42] (see also the later papers by Bagrov[43] and Gruza[44]), to be specific let us consider a scheme in which the predicted phenomenon Φ allows a finite number n of possible outcomes or phases $\Phi_1, \Phi_2, \ldots, \Phi_n$ (in the particular case of the *alternative* forecast only two phases are considered: the occurrence or nonoccurrence of a phenomenon; this case was studied, e.g., in the earlier papers by Omshansky[45] and Obukhov[46]). In particular, such phases may be the intervals of values of the forecasted quantity (the air temperature, the amount of precipitation, the height of a flood, etc.) or the dates when phenomena occur (the temperature crossing zero, the formation or melting of river ice, etc.).

Let a forecast be given in the form of one of the texts $\Pi_1, \Pi_2, \ldots, \Pi_n$, with text Π_i being the prediction that the phenomenon will occur in phase Φ_i. It is possible to consider that the occurrence of the phase Φ_α and the forecaster's choosing text Π_β are a pair of conjugate random events with a probability $p_{\alpha\beta} = P(\Phi_\alpha, \Pi_\beta)$. A table of the $n \times n$ probabilities $p_{\alpha\beta}$ is the most complete characterization of the natural variability of a phenomenon (the quantities $p_{\alpha\cdot} = \Sigma_\beta p_{\alpha\cdot\beta}$ are the *climatological probabilities* of the phases Φ_α, and

$$H(\Phi) = - \sum_\alpha p_{\alpha\cdot} \log p_{\alpha\cdot}. \tag{24.1}$$

is the entropy of the phenomenon) as well as the reliability of the forecast method (the quantities

$$r_{i\beta} = P(\Pi_\beta | \Phi_i) = \frac{p_{i\beta}}{p_{i\cdot}} \tag{24.2}$$

are the *probabilities of errors* in the forecast, according to which the different methods of forecasting must be compared). If the forecast

Π_j is given, then the occurrence of the phases Φ_α should not be expected with the climatological probabilities $p_\alpha.$, but with the probabilities

$$q_{\alpha j} = p(\Phi_\alpha | \Pi_j) = \frac{p_{\alpha j}}{p._j}, \tag{24.3}$$

where $p._j = \Sigma_\alpha p_{\alpha j}$. In this substitution of absolute probabilities $p_\alpha.$ for the conditional probabilities $q_{\alpha j}$ lies the significance of the forecaster's work. The fact that a forecast Π_j does not uniquely determine which phase of the phenomenon will occur, but only gives a distribution of probabilities $q_{\alpha j}$ for the phases Φ_α, is an expression of the unreliability of the prognostic method. Indicating the distribution of probabilities $q_{\alpha j}$ along with the forecast Π_j changes the forecast from a *categorical* to a more complete *probabilistic* one. Under the condition Π_j the entropy of the phenomenon *decreases* to a value

$$H(\Phi|\Pi_j) = - \sum_\alpha q_{\alpha j} \log q_{\alpha j}. \tag{24.4}$$

If the average conditional entropy

$$H(\Phi|\Pi) = \sum_\beta p._\beta H(\Phi|\Pi_\beta) \tag{24.5}$$

is introduced, then the magnitude of

$$I(\Phi|\Pi) = H(\Phi) - H(\Phi|\Pi) \tag{24.6}$$

will be a measure of the *amount of information* about phenomena that is contained in forecasts according to a given method [it must be compared to the entropy $H(\Phi)$].

A specific user's economic dependence on a phenomenon Φ is described by a table of $n \times n$ numbers $S_{ij} = S(\Phi_i, \Pi_j)$, his profits or losses in cases when he has planned on the occurrence of an event in phase Φ_j (i.e., has trusted in the forecast Π_j) and has carried out the corresponding preliminary measures, but the phenomenon then occurs in phase Φ_i. Without knowledge of such a table the full-valued economic use of any forecasts is impossible (however, as far as we know, such tables have still not been compiled, even for the main users served by hydrometeorological organizations; in particular, this makes it difficult to estimate the economic justification of public expenditures for the maintenance of the hydrometeorological services).

If the user relies fully on forecasts according to a given method, then in cases with a forecast Π_j he will receive an average profit or loss of

$$S_{\cdot j} = \sum_\alpha q_{\alpha j} S_{\alpha j},$$

and the average effect of all forecasts will be equal to

$$S = \sum_\beta p_{\cdot \beta} S_{\cdot \beta} = \sum_{\alpha\beta} p_{\alpha\beta} S_{\alpha\beta}. \tag{24.7}$$

If the user does not receive any forecasts but knows the climatological probabilities p_α. of the phases Φ_α, then it is most advantageous[41] for him to plan always on the occurrence of the same phase—the one to which corresponds the largest of the numbers

$$S_{\cdot j}^* = \sum_\alpha p_\alpha \cdot S_{\alpha j};$$

this will then be his average income. On the other hand, in obtaining the forecasts $\Pi_1, \Pi_2, \ldots, \Pi_n$ the user's *optimum strategy* consists in the following: In cases with a forecast Π_j he should plan on the occurrence of the phase Φ_{mj} corresponding to the largest of the numbers $S_{\cdot 1}^{(j)}, S_{\cdot 2}^{(j)}, \ldots, S_{\cdot n}^{(j)}$, which are defined by the formula

$$S_{\cdot \beta}^{(j)} = \sum_\alpha q_{\alpha j} S_{\alpha\beta}.$$

In other words, this user should replace the forecasts $\Pi_1, \Pi_2, \ldots,$ Π_n by $\Pi_{m_1}, \Pi_{m_2}, \ldots, \Pi_{m_n}$ (some of the latter may coincide). The average effect obtained by the user will then be the greatest; it can be calculated from the formula

$$S_{\max} = \sum_{\beta} p_{\cdot\beta} S_{\cdot m_\beta}{}^{(\beta)}. \tag{24.8}$$

This quantity can turn out considerably larger than (24.7).

Finally, studying the tables of S_{ij} and the quantities $S_{\cdot j}^*$, $S_{\cdot j}$, and $S_{\cdot m_\beta}{}^{(\beta)}$ can facilitate the planning of constant insurance arrangements that would change the table of S_{ij} so as to decrease the user's economic dependence on the predicted elemental weather phenomena and increase his average income.

References

Chapter 1

1. A. Kh. Khrgian, *Ocherki razvitiya meteorologii* [Outlines of the development of meteorology], Gidrometizdat, 1948.
2. I. A. Kibel', *Izv. Akad. Nauk SSSR, Ser. Geogr. i Geofiz.* [Bull. Acad. Sci USSR, Geograph. and Geophys. Series] **4**, 627 (1940).
3. A. M. Obukhov, ibid. **13**, 281 (1949).
4. J. G. Charney, *Geofys. Publikasjoner, Norske Videnskaps-Akad. Oslo* **17**, 17 (1948).
5. J. G. Charney, *J. Meteorol.* **6**, 371 (1949).
6. J. G. Charney, R. Fjørtoft, J. von Neumann, *Tellus* **2**, 237 (1950).
7. I. A. Kibel, *Introduction to the Hydrodynamical Methods of Short Period Weather Forecasting* (trans. from Russian), Pergamon, New York, 1963.
8. P. D. Thompson, *Numerical Weather Analysis and Prediction*, Macmillan, New York, 1961.
9. P. N. Belov, *Prakticheskiye metody chislennogo prognoza pogody* [Practical methods of numerical weather forecasting], 2d ed., Gidrometizdat, 1967.
10. G. I. Marchuk, *Chislennye metody v prognoze pogody* [Numerical methods in weather forecasting], Gidrometizdat, 1967.
11. E. N. Blinova, *Dokl. Akad. Nauk SSSR* [Rept. Acad. Sci. USSR] **39**, 284 (1943).
12. A. S. Monin, *Meteorol. i Gidrol.* [Meteorol. and Hydrol.], No. 8, 43 (1963).
13. E. P. Borisenkov, *Tr. Arkticheskogo i Antarkt. Inst.* [Trans. Arctic and Antarctic Inst.] **253**, 109 (1963).
14. P. R. Pisharoty, *Final Rept. Gen. Circ. Proj.*, AF19 (122)—48 (1955).
15. G. V. Gruza, *Integral'nye kharakteristiki obshchey tsirkulyatsii atmosfery* [Integral characteristics of the general circulation of the atmosphere], Gidrometizdat, 1965.
16. E. Palmén, *Geofis. Pura Appl.* **49**, 167 (1961).
17. E. Palmén, in *The Atmosphere and the Sea in Motion*, ed. B. Bolin, Rockefeller Inst. Press in assoc. with the Oxford Univ. Press, New York, 1959, pp. 212–224.
18. D. Brunt, *Physical and Dynamical Meteorology*, 2d ed., Cambridge Univ. Press, New York, 1939.
19. L. F. Richardson, *Proc. Roy. Soc. (London), Ser. A* **110**, 708 (1926).
20. W. Rudloff, *Wetterlotse* **14**, 183 (1962).
21. R. Neik, *Nat. Hist.*, No. 50 (1960).
22. A. M. Obukhov, *Dokl. Akad. Nauk SSSR* **32**, 22 (1941); *Izv. Akad. Nauk SSSR, Ser. Geogr. i Geofiz.* **5**, 453 (1941).
23. A. M. Obukhov, *Izv. Akad. Nauk SSSR, Ser. Geogr. i Geofiz.* **13**, 58 (1949).
24. G. S. Golitsyn, *Izv. Akad. Nauk SSSR, Ser. Geofiz.*, 1253 (1964).
25. V. N. Kolesnikova and A. S. Monin, *Izv. Akad. Nauk SSSR, Fiz. Atmosfery i Okeana* [Phys. of the Atmosphere and Ocean] **1**, 653 (1965).
26. J. van der Hoven, *J. Meteorol.* **14**, 160 (1957).
27. C. G. Rossby, in *The Atmospheres of the Earth and Planets*, Univ. of Chicago Press, Chicago, 1948, p. 25.

28. E. N. Lorenz, *Tellus* **5**, 238 (1953).
29. B. L. Dzerdzeyevsky and A. S. Monin, *Izv. Akad. Nauk. SSSR, Ser. Geofiz.*, 562 (1954).
30. A. S. Monin, *Izv. Akad. Nauk SSSR, Ser. Geofiz.*, 452 (1956).
31. A. S. Monin, *Proc. Symp. Time Series Analysis*, Wiley, New York, 1963, pp. 144–151.
32. V. N. Kolesnikova and A. S. Monin, *Izv. Akad. Nauk SSSR, Fiz. Atmosfery i Okeana* **2**, 113 (1966).
33. B. L. Dzerdzeyevesky, in *A. I. Voyeykov i problemy sovremennoy klimatologii* [A. I. Voyeykov and the problems of modern climatology], Gidrometizdat, 1956, pp. 109–122.
34. R. A. Craig and H. C. Willett, in *Compendium of Meteorology*, ed. T. F. Malone, Am. Meteorol. Soc., Boston, 1951, pp. 379–390.
35. G. C. Simpson, *Nature* **141**, 591 (1938).

Chapter 2

1. A. S. Monin, *Meteorol. i Gidrol.*, No. 8, 43 (1963).
2. C. G. Rossby, *Quart. J. Roy. Meteorol. Soc.* **66**, Suppl., 68 (1940).
3. H. Ertel, *Meteorol. Z.* **59**, 277 (1942).
4. J. G. Charney, *Geofys. Publikasjoner, Norske Videnskaps-Akad. Oslo* **17**, 17 (1948).
5. H. Arakawa, *Geofis. Pura Appl.* **18**, 159 (1950).
6. A. M. Obukhov, *Dokl. Akad. Nauk SSSR* **145**, 1239 (1962).
7. A. M. Obukhov, *Meteorol. i Gidrol.*, No. 2, 3 (1964).
8. G. H. Hollmann, *Arch. Meteorol., Geophys. Bioklimatol., Ser. A* **14**, 1 (1964).
9. E. N. Lorenz, *Tellus* **7**, 157 (1955).
10. E. N. Lorenz, *Dynamics of Climate*, Pergamon, London, 1960, pp. 86–92.
11. E. N. Lorenz, *Tellus* **12**, 364 (1960).
12. A. S. Monin and A. M. Obukhov, *Dokl. Akad. Nauk SSSR* **122**, 58 (1958); *Izv. Akad. Nauk SSSR, Ser. Geofiz.*, 1360 (1958); *Tellus* **11**, 159 (1959).
13. A. S. Monin, *Izv. Akad. Nauk SSSR, Ser. Geofiz.*, 497 (1958).
14. C. Eckart, *Hydrodynamics of Oceans and Atmospheres*, Pergamon, London, 1960.
15. A. M. Obukhov, *Izv. Akad. Nauk SSSR, Ser. Geogr. i Geofiz.* **13**, 281 (1949).
16. L. A. Diky, *Izv. Akad. Nauk SSSR, Ser. Geofiz.*, 1186 (1959).
17. C. G. Rossby, *J. Marine Res.* **2**, 38 (1939).
18. S. S. Hough, *Phil. Trans. Roy. Soc. London, Ser. A* **189**, 201 (1897); **191**, 139 (1898).
19. A. E. Love, *Proc. London Math. Soc., Ser. 2* **12**, 813 (1913).
20. N. E. Kochin, *Sobr. soch.* [Coll. works], vol. 1, Moscow and Leningrad, 1949.
21. C. L. Pekeris, *Proc. Roy. Soc. (London), Ser. A* **158**, 650 (1937).
22. W. Kertz, *Ann. Meteorol.*, Beiheft (1951).

23. M. Seibert, *Advan. Geophys.* **7** (1961).
24. A. M. Yaglom, *Izv. Akad. Nauk SSSR, Ser. Geofiz.*, 346 (1963).
25. B. Haurwitz, *J. Marine Res.* **3**, 254 (1940).
26. E. N. Blinova, *Dokl. Akad. Nauk SSSR* **39**, 284 (1943).
27. L. A. Diky, *Izv. Akad. Nauk SSSR, Ser. Geofiz.*, 756 (1961).
28. L. A. Diky, *Izv. Akad. Nauk SSSR, Fiz. Atmosfery i Okeana* **1**, 469 (1965).
29. C. G. Rossby, *J. Marine Res.* **1**, 239 (1938).
30. A. Cahn, *J. Meteorol.* **2**, 113 (1945).
31. B. Bolin, *Tellus* **5**, 373 (1953).
32. I. A. Kibel', *Dokl. Akad. Nauk SSSR* **104**, 60 (1955).
33. G. Veronis, *Deep-Sea Res.* **3**, 157 (1956).
34. J. E. Fjelstad, *Geofys. Publikasjoner Norske Videnskaps-Akad. Oslo* **20**, 1 (1958).
35. N. A. Phillips, *Rev. Geophys.* **1**, 123 (1963).
36. A. S. Monin, *Izv. Akad. Nauk SSSR, Ser. Geofiz.*, 602 (1961).
37. J. G. Charney, *Proc. Int. Symp. Numerical Weather Prediction, Tokyo,* Meteorol. Soc. Japan, Tokyo, 1962, p. 131.
38. J. G. Charney, and M. E. Stern, *J. Atmos. Sci.* **19**, 159 (1962).
39. B. L. Gavrilin, *Izv. Akad. Nauk SSSR, Fiz. Atmosfery i Okeana* **1**, 557 (1965).
40. A. S. Chaplygina, *Tr. Geofiz. Inst. Akad. Nauk SSSR, Sb. Statei* [Trans. Geophys. Inst. Acad. Sci. USSR, Collection of Articles], No. 33 (160), 60 (1956).
41. I. A. Kibel', *Izv. Akad. Nauk SSSR, Ser. Geogr. i Geofiz.* **4**, 627 (1940).
42. A. S. Monin, *Izv. Akad. Nauk SSSR, Ser. Geofiz.*, 76 (1952).
43. J. G. Charney, *J. Meteorol.* **6**, 371 (1949).
44. H. E. Ertel, *Meteorol. Z.* **58**, 77 (1941).
45. J. G. Charney, R. Fjørtoft, and J. von Neumann, *Tellus* **2**, 237 (1950).
46. A. M. Obukhov and A. S. Chaplygina, *Tr. Inst. Fiz. Atmosfery Akad. Nauk SSSR* [Trans. Inst. Atmos. Phys. Acad. Sci. USSR], No. 2, 23 (1958).
47. N. I. Buleyev, G. I. Marchuk, *Tr. Inst. Fiz. Atmosfery Akad. Nauk SSSR*, No. 2, 66 (1958).
48. K. Hinkelmann, *Tellus* **5**, 499 (1953).
49. H. L. Kuo, *Tellus* **8**, 373 (1956).
50. B. Bolin, *Tellus* **7**, 27 (1955).
51. P. D. Thompson, *J. Meteorol.* **13**, 251 (1956).
52. J. G. Charney, *Tellus* **7**, 22 (1955).
53. J. Smagorinsky, *Monthly Weather Rev.* **86**, 457 (1958).
54. K. Hinkelmann, in *The Atmosphere and the Sea in Motion,* ed. B. Bolin, Rockefeller Inst. Press in assoc. with the Oxford Univ. Press, New York, 1959.
55. N. A. Phillips, *Monthly Weather Rev.* **87**, 333 (1959).
56. N. I. Buleyev and G. I. Marchuk, *Tr. Vses. Nauchn. Meteorol. Soveshch.* [Trans. All-Union Sci. Meteorol. Conf.], vol. 2, Leningrad, 1963.
57. G. I. Marchuk, *Dokl. Akad. Nauk SSSR* **155**, 1062 (1964).
58. G. I. Marchuk, ibid. **156**, 308 (1964).
59. G. I. Marchuk, ibid. **156**, 810 (1964).

60. G. I. Marchuk, *Chislennye metody v prognoze pogody* [Numerical methods in weather forecasting], Gidrometizdat, 1967.
61. K. Hinkelmann, *Tellus* **3**, 285 (1951).
62. A. M. Obukhov, *Tr. Geofiz. Inst. Akad. Nauk SSSR, Sb. Statei*, No. 24 (151), 3 (1954).
63. A. M. Obukhov, *Izv. Akad. Nauk SSSR, Ser. Geofiz.*, 432 (1960).
64. L. V. Rukhovets, *Izv. Akad. Nauk SSSR, Ser. Geofiz.*, 626 (1963).
65. A. A. Fukuoka, *Geophys. Mag.*, 177 (1951).
66. E. N. Lorenz, Sci. Rept. MIT NI, contract AF19 (604), 1956.
67. R. M. White, D. S. Cooley, R. S. Derby, and F. A. Seaver, *J. Meteorol.* **15**, 426 (1958).
68. N. A. Bagrov, *Tr. Tsentr. Inst. Progn.* [Trans. Cent. Forecast Inst.] **74**, 3 (1959).
69. J. Holmström, *Tellus* **15**, 127 (1963).
70. A. S. Monin and A. M. Obukhov, in *Dinamika krupnomasshtabnykh atmosfernykh protsessov* [The dynamics of large-scale atmospheric processes], Nauka, 1967, pp. 194–200.
71. B. L. Gavrilin, *Izv. Akad. Nauk SSSR, Fiz. Atmosfery i Okeana* **1**, 8 (1965).
72. A. M. Obukhov, *Izv. Akad. Nauk SSSR, Ser. Geofiz.*, 1133 (1957).
73. S. K. Godunov and V. S. Ryaben'ky, *Vvedeniye v teoriyu raznostnykh skhem* [Introduction to the theory of difference schemes], Fizmatgiz, 1962.
74. R. Courant, K. Friedrichs, and H. Lewy, *Math. Ann.*, 100 (1928).
75. N. A. Phillips, in *The Atmosphere and the Sea in Motion*, ed. B. Bolin, Rockefeller Inst. Press in assoc. with the Oxford Univ. Press, New York, 1959, p. 501.
76. J. Smagorinsky, *Monthly Weather Rev.* **91**, 99 (1963).
77. C. E. Leith, in *Methods in Computational Physics*, ed. B. Alder, S. Fernbach, and M. Rotenberg, *Vol. 4: Applications in Hydrodynamics*, Academic Press, New York, 1965, p. 769.
78. Y. Mintz, *WMO-IUGG Symp. Res. Develop. Aspects of Long-Range Forecasting*, WMO Tech. Note No. 66, 1965, p. 141.
79. D. Lilly, *Monthly Weather Rev.* **93**, 11 (1965).
80. F. G. Schuman, *Proc. Int. Symp. Numerical Weather Prediction, Tokyo*, Meteorol. Soc. Japan, Tokyo, 1962, p. 85.
81. G. Platzman, *J. Meteorol.* **17**, 635 (1960).
82. G. Fisher, *Monthly Weather Rev.* **93**, 1 (1965).
83. E. Knighting, *Quart. J. Roy. Meteorol. Soc.* **86**, 318 (1960).
84. W. Lewis, *Monthly Weather Rev.* **85**, 297 (1957).
85. Y. Nabeshima, *Proc. Int. Symp. Numerical Weather Prediction, Tokyo*, Meteorol. Soc. Japan, Tokyo, 1962, p. 265.
86. M. E. Shvets, *Tr. Gl. Geofiz. Observ.* [Trans. Main Geophys. Obs.] **81**, 3 (1959).
87. P. K. Dushkin, E. G. Lomonosov, and Yu. N. Lunin, *Meteorol. i Gidrol.*, No. 12, 3 (1960).
88. J. Vederman, *Monthly Weather Rev.* **89**, 243 (1961).
89. V. V. Ovsyannikov, *Tr. Tsentr. Inst. Progn.* **81**, 92 (1961).

90. V. V. Ovsyannikov, *Meteorol. i Gidrol.*, No. 3, 67 (1962).
91. L. A. Kuznetsov, ibid., No. 12, 9 (1962).
92. K. Pedersen, *Geofys. Publikasjoner, Norske Videnskaps-Akad. Oslo* **25**, 25 (1963).
93. B. D. Uspensky, *Meteorol. i Gidrol.*, No. 2, 3 (1965).
94. N. A. Bagrov, ibid., No. 12, 20 (1965).
95. H. H. Lamb, *Quart. J. Roy. Meteorol. Soc.* **81**, 172 (1955).
96. A. I. Wagner, *Bull. Am. Meteorol. Soc.* **38**, 584 (1957).
97. J. Smagorinsky and G. O. Collins, *Monthly Weather Rev.* **83**, 53 (1955).
98. J. Smagorinsky, *Ber. des Deut. Wetterdienstes* **5**, 82 (1957).
99. S. I. Smebye, *J. Meteorol.* **15**, 547 (1958).
100. J. Smagorinsky, *Physics of Precipitation*, Monograph No. 5, Am. Geophys. Union, 1960, p. 71.
101. V. S. Antonov, *Meteorol. i Gidrol.*, No. 9, 54 (1963).
102. L. T. Matveyev, *Tr. Arkticheskogo i Antarkt. Inst.*, No. 228, 14 (1959).
103. L. T. Matveyev, *Izv. Akad. Nauk SSSR, Ser. Geofiz.*, 130 (1961).
104. L. T. Matveyev, *Meteorol. i Gidrol.*, No. 2, 1 (1962).
105. L. T. Matveyev, *Kosmich Issled., Acad. Nauk SSSR* [Cosmic Res., Acad. Sci. USSR] **2**, 109 (1964).
106. L. T. Matveyev, *Tellus* **16**, 139 (1964).
107. L. T. Matveyev, *Osnovy obshchey meteorologii (Fizika atmosfery)* [Fundamentals of general meteorology (Physics of the atmosphere)], Gidrometizdat, 1965.
108. Yu. G. Lushev and L. T. Matveyev, *Izv. Akad. Nauk SSSR, Fiz. Atmosfery i Okeana* **2**, 3 (1966).
109. L. P. Bykova and L. T. Matveyev, ibid. **2**, 905 (1966).
110. E. M. Feygel'son and N. G. Frolova, ibid. **1**, 241 (1965).
111. S. S. Zilitinkevich and D. L. Laykhtman, *Dokl. Akad. Nauk SSSR* **156**, 1079 (1964).
112. M. E. Shvets, *Izv. Akad. Nauk SSSR, Ser. Geofiz.*, 547 (1955).
113. S. L. Lebedev, *Izv. Akad. Nauk SSSR, Fiz. Atmosfery i Okeana* **1**, 456 (1965).
114. S. L. Lebedev, ibid. **2**, 14 (1966).
115. G. I. Marchuk, *Izv. Akad. Nauk SSSR, Ser. Geofiz.*, 754 (1964).
116. N. H. Fletcher, *The Physics of Rainclouds,* Cambridge Univ. Press, New York, 1962.
117. M. V. Buykov, *Tr. Vses. Meteorol. Soveshch. v Leningrade* [Trans. All-Union Meteorol. Conf. in Leningrad] **5**, 122 (1963).
118. Yu. V. Shulepov and M. V. Buykov, *Izv. Akad. Nauk SSSR, Fiz. Atmosfery i Okeana* **1**, 248 (1965).
119. Yu. V. Shulepov and M. V. Buykov, ibid. **1**, 353 (1965).
120. H. Riehl, *Science* **141**, 1001 (1963).
121. H. Riehl, *Tropical Meteorology*, McGraw-Hill, New York, 1954.
122. Z. M. Tiron, *Uragany* [Hurricanes], Gidrometizdat, 1964.
123. Simada Kandzi, *Tenki* **9**, 164 (1962).
124. R. Fett, *Monthly Weather Rev.* **92**, 43 (1964).

125. E. C. Hill, *Mariners Weather Log* **6**, 153 (1962).
126. M. A. Estoque, *Geofis. Int.* **3**, 133 (1963).
127. G. K. Morikawa, *J. Meteorol.* **17**, 148 (1960).
128. G. K. Morikawa, *Proc. Int. Symp. Numerical Weather Prediction, Tokyo,* Meteorol. Soc. Japan, Tokyo, 1962, p. 349.
129. L. Onsager, *Nuovo Cimento, Suppl.* **6**, 3 (1949).
130. E. Fermi, J. Pasta, and S. Ulam, *Los Alamos Sci. Lab. Rept. LA–1557,* 1953.
131. J. Pasta and S. Ulam, *Los Alamos Sci. Lab. Rept. LA–1940,* 1955.
132. S. Ulam, *Proc. Symp. Appl. Math.* **13**, 247–258 (1962).
133. J. G. Charney, *Proc. Symp. Appl. Math.* **15**, 289 (1963).

Chapter 3

1. N. E. Kochin, *Tr. Gl. Geofiz. Observ.* **4**, (1936).
2. E. N. Blinova, *Dokl. Akad. Nauk SSSR* **39**, 284 (1943).
3. E. N. Blinova, *Izv. Akad. Nauk SSSR, Ser. Geogr. i Geofiz.* **11**, 3 (1947).
4. S. A. Mashkovich, *Tr. Tsentr. Inst. Progn.* **78**, 5 (1958).
5. E. N. Blinova, *Tr. Inst. Fiz. Atmosfery Akad. Nauk SSSR,* No. 2, 5 (1958).
6. E. N. Blinova and G. I. Marchuk, ibid., No. 2, 105 (1958).
7. E. N. Blinova and I. A. Kibel', *Tellus* **9**, 447 (1957).
8. E. N. Blinova, *Tr. Mirov. Meteorol. Tsentra* [Trans. Mirov Meteorol. Center] **2**, 3 (1964).
9. E. N. Blinova, *Izv. Akad. Nauk SSSR, Ser. Geofiz.,* 110 (1964).
10. E. N. Blinova, *WMO-IUGG Symp. Res. Develop. Aspects of Long-Range Forecasting,* WMO Tech. Note No. 66, 1965, p. 63.
11. E. N. Blinova, *Tr. Mirov. Meteorol. Tsentra* **5**, 14 (1965).
12. E. N. Blinova, in *Dinamika krupnomasshtabnykh atmosfernykh protsessov* [The dynamics of large-scale atmospheric processes], Nauka, 1967, p. 15.
13. Global Atmospheric Research Programme (GARP), Rept. of the Study Conf. Held at Stockholm 28 June–11 July 1967.
14. A. S. Monin, *Meteorol. i Gidrol.,* No. 8, 43 (1963).
15. J. Charney, in *Dinamika krupnomasshtabnykh atmosfernykh protsessov* [The dynamics of large-scale atmospheric processes], Nauka, 1967, p. 21.
16. E. Palmén, *J. Meteorol.* **5**, 20 (1948).
17. E. R. Reiter, *Meteorologie der Strahlstrome,* Springer, Vienna, 1961.
18. J. C. Sadler, *Bull. Am. Meteorol. Soc.* **46**, 118 (1965).
19. H. Riehl, *Tropical Meteorology,* McGraw-Hill, New York, 1954.
20. J. G. Charney, *J. Atmos. Sci.* **20**, 607 (1963).
21. J. G. Charney, *Geofis. Int.* **3**, 69 (1963).
22. J. G. Charney, *Proc. Seminar on NJVP, Travelers Res. Center, Hartford, Conn., 1966,* vol. 1.
23. L. I. Litvinenko, *Meteorol. i Gidrol.,* No. 6, 29 (1965).

24. L. I. Litvinenko, *Tr. Tsentr. Inst. Progn.* **146**, 29 (1965).
25. R. J. Reed, in *Dinamika krupnomasshtabnykh atmosfernykh protsessov* [The dynamics of large-scale atmospheric processes], Nauka, 1967, p. 393.
26. M. S. Malkevich, A. S. Monin, and G. V. Rozenberg, *Izv. Akad. Nauk SSSR, Fiz. Atmosfery i Okeana,* 394 (1964).
27. L. Kletter, *Phys. Bl.* **19**, 116 (1963).
28. G. Lariviere, *Sci. Progr., Nat. (Paris),* 489 (1963).
29. K. Ya. Kondrat'yev, *Meteorologicheskiye issledovaniya s pomoshchyu raket i sputnikov* [Meteorological studies with the aid of rockets and satellites], Gidrometizdat, 1963.
30. K. Ya. Kondrat'yev, *Meteorologicheskiye sputniki* [Weather satellites], Gidrometizdat, 1963.
31. A. Arking, *Science* **143**, 569 (1964).
32. J. B. Jones, *Monthly Weather Rev.* **89**, 383 (1961).
33. L. F. Hubert, *Monthly Weather Rev.* **89**, 229 (1961).
34. J. S. Winston, *Ann. N.Y. Acad. Sci.* **93**, 775 (1962).
35. A. F. Krueger and S. Fritz, *Tellus* **13**, 1 (1961).
36. C. H. B. Priestley, *Tellus* **14**, 123 (1962).
37. A. Icart, *Sci. et Voyages,* No. 210, 15 (1963).
38. K. Ya. Kondrat'yev, E. P. Borisenkov, and A. A. Morozkin, *Prakticheskoye ispol'zovaniye dannykh meteorologicheskykh sputnikov* [The practical use of weather satellite data], Gidrometizdat, 1966.
39. W. Nordberg, W. R. Bandeen, G. Warnecke, and V. Kunde, in *Space Research V,* North-Holland, Amsterdam, 1965.
40. W. Nordberg, A. W. McCulloch, L. L. Foshee, and W. R. Bandeen, *Bull. Am. Meteorol. Soc.* **47**, 857 (1966).
41. D. Q. Wark and H. C. Fleming, *Monthly Weather Rev.* **94**, 351 (1966).
42. A. I. Kliore, G. S. Levy, V. R. Eschelman, C. Fjelbo, and F. Drake, *Science* **149**, 1243 (1965).
43. W. H. Gregory, *Aviat. Week Space Technol.* **82**, 54 (1965).
44. S. F. Singer, *Int. Sci. Technol.,* No. 36, 30 (1964).
45. A. M. Obukhov, *Dokl. Akad. Nauk SSSR* **145**, 1239 (1962).
46. B. L. Gavrilin, *Izv. Akad. Nauk SSSR, Fiz. Atmosfery i Okeana* **1**, 557 (1965).
47. I. A. Kibel, *Introduction to the Hydrodynamical Methods of Short Period Weather Forecasting* (trans. from Russian), Pergamon, New York, 1963.
48. A. S. Monin, *Dokl. Akad. Nauk SSSR* **175**, 819 (1967).
49. S. Manabe and F. Möller, *Monthly Weather Rev.* **89**, 503 (1961).
50. S. Manabe and R. Strickler, *J. Atmos. Sci.* **21**, 361 (1964).
51. G. I. Marchuk, *Izv. Akad. Nauk SSSR, Ser. Geofiz.,* 754 (1964).
52. G. I. Marchuk, *Chislennye metody v prognoze pogody* [Numerical methods in weather forecasting], Gidrometizdat, 1967.
53. A. A. Nichiporovich, *Svetovoye i uglerodnoye pitaniye rasteny (fotosintez)* [Plant feeding by light and carbon (photosynthesis)], *Izd. Acad. Nauk SSSR,* 1955.
54. H. W. Chapman, L. S. Gleason, and W. E. Loomis, *Plant Physiol.* **29**, 500 (1954).

55. B. Bolin and C. D. Keeling *J. Geophys. Res.* **68**, 3899 (1963).
56. K. S. Shifrin and O. A. Avaste, *Tr. Inst. Fiz. i Astron., Akad. Nauk Est. SSR* [Trans. Inst. of Phys. and Astron., Acad. Sci. Estonian SSR], *Issled. po Fiz. Atmosfery* [Studies in Atmos. Phys.], No. 2, (1960).
57. Kh. Yu. Nylisk, ibid., No. 4, (1963).
58. E. M. Feygel'son, *Radiatsionnye protsessy v sloistoobraznykh oblakakh* [Radiational processes in stratus clouds], Nauka, 1964.
59. E. M. Feygel'son, *Izv. Akad. Nauk SSSR, Ser. Geofiz.*, 1539 (1964).
60. G. V. Rozenberg, *Izv. Akad. Nauk SSSR, Fiz. Atmosfery i Okeana* **3**, 936 (1967).
61. A. S. Monin, in *Nauka i chelovechestvo* [Science and man], Znaniye, 1964, pp. 192–204.
62. B. L. Gavrilin and A. S. Monin, *Dokl. Akad. Nauk SSSR* **176**, 822 (1967).
63. W. Lewis, *Monthly Weather Rev.* **85**, 297 (1957).
64. R. V. Ozmidov and N. I. Popov, *Izv. Akad. Nauk SSSR, Fiz. Atmosfery i Okeana* **2**, 183 (1966).
65. N. A. Phillips, *Quart. J. Roy. Meteorol. Soc.* **82**, 123 (1956).
66. J. Smagorinsky, *Monthly Weather Rev.* **91**, 99 (1963).
67. J. Smagorinsky, *Quart. J. Roy. Meteorol. Soc.* **90**, 1 (1964).
68. J. Smagorinsky, S. Manabe, and J. L. Holloway, *Monthly Weather Rev.* **93**, 727 (1965).
69. J. Smagorinsky, *WMO-IUGG Symp. Res. Develop. Aspects of Long-Range Forecasting*, WMO Tech. Note No. 66, 1965, p. 131.
70. S. Manabe, J. Smagorinsky, and R. F. Strickler, *Monthly Weather Rev.* **93**, 769 (1965).
71. J. Smagorinsky et al., in *Dinamika krupnomasshtabnykh atmosfernykh protsessov* [The dynamics of large-scale atmospheric processes], Nauka, 1967, p. 70.
72. S. Manabe and J. Smagorinsky, *Monthly Weather Rev.* **95**, 155 (1967).
73. Y. Mintz, *WMO-IUGG Symp. Res. Develop. Aspects of Long-Range Forecasting*, WMO Tech. Note No. 66, 1965, p. 141.
74. Y. Mintz and A. Arakawa, *Trans. Am. Geophys. Union* **44**, 53 (1963).
75. Y. Mintz and A. Arakawa, *Proc. Symp. Arctic Heat Budget and Atmos. Circulation*, Mem. RM–5233–NSF, 1966, p. 369.
76. Y. Mintz, in *Dinamika krupnomasshtabnykh atmosfernykh protsessov* [The dynamics of large-scale atmospheric processes], Nauka, 1967, p. 138.
77. F. Mesinger, in *Dinamika krupnomasshtabnykh atmosfernykh protsessov* [The dynamics of large-scale atmospheric processes], Nauka, 1967, p. 139.
78. C. E. Leith, in *Methods in Computational Physics,* ed. B. Alder, S. Fernbach, and M. Rotenberg, *Vol. 4: Applications in Hydrodynamics,* Academic Press, New York, 1965, p. 769.
79. C. E. Leith, *WMO-IUGG Symp. Res. Develop. Aspects of Long-Range Forecasting*, WMO Tech. Note No. 66, 1965, p. 168.
80. C. E. Leith, *Proc. Symp. Arctic Heat Budget and Atmos. Circulation*, Mem. RM–5233–NSF, 1966, p. 371.

81. C. E. Leith, in *Dinamika krupnomasshtabnykh atmosfernykh protsessov* [The dynamics of large-scale atmospheric processes], Nauka, 1967, p. 134.
82. S. Matsumoto, *Proc. Int. Symp. Numerical Weather Prediction, Tokyo*, Meteorol. Soc. Japan, Tokyo, 1962, p. 557.
83. Chen Jung-san, *Acta Meteorol. Sinica* **34**, 443 (1964).
84. Chen Jung-san, *Sci. Sinica (Peking)* **14**, 246 (1965).
85. A. Kasahara and W. M. Washington, *Proc. Symp. Arctic Heat Budget and Atmos. Circulation*, Mem. RM–5233–NSF, 1966, p. 401.
86. K. Gambo, in *Dinamika krupnomasshtabnykh atmosfernykh protsessov* [The dynamics of large-scale atmospheric processes], Nauka, 1967, p. 152.
87. J. Adem, *Tellus* **14**, 102 (1962).
88. J. Adem, *Monthly Weather Rev.* **91**, 375 (1963).
89. J. Adem, *Geofis. Int.* **4**, 1964.
90. J. Adem, *Monthly Weather Rev.* **92**, 91 (1964).
91. J. Adem, *WMO-IUGG Symp. Res. Develop. Aspects of Long-Range Forecasting*, WMO Tech. Note No. 66, 1965, p. 138.
92. B. L. Gavrilin, *Izv. Akad. Nauk SSSR, Fiz. Atmosfery i Okeana* **1**, 1229 (1965).
93. A. S. Sarkisyan, *Dokl. Akad. Nauk SSSR* **134**, 1339 (1960).
94. A. S. Sarkisyan, *Izv. Akad. Nauk SSSR, Ser. Geofiz.*, 1396 (1961).
95. A. S. Sarkisyan, *Okeanologiya* **2**, No. 3, (1962).
96. Yu. K. Garmatyuk and A. S. Sarkisyan, *Izv. Akad. Nauk SSSR, Fiz. Atmosfery i Okeana* **1**, 313 (1965).
97. A. S. Sarkisyan, *Osnovy teorii i raschet okeanicheskykh techeny* [Fundamentals of the theory and calculation of ocean currents], Gidrometizdat, 1966.
98. K. Bryan, in *Methods in Computational Physics*, ed. B. Alder, S. Fernbach, and M. Rotenberg, *Vol. 4: Applications in Hydrodynamics*, Academic Press, New York, 1965.
99. K. Bryan and M. D. Cox, *Tellus* **19**, 54 (1967).
100. K. Takahashi, A. Katayama, and T. Asakura, *J. Meteorol. Soc. Japan* **38**, 175 (1960).
101. Y. Mintz, *Bull. Am. Meteorol. Soc.* **35**, 208 (1954).
102. A. H. Oort, *Monthly Weather Rev.* **92**, 483 (1964).
103. R. M. White and B. Saltzman, *Tellus* **8**, 357 (1956).
104. A. F. Krueger, J. S. Winston, and D. Haynes, *Monthly Weather Rev.* **93**, 227 (1965).
105. A. I. Wiin-Nielsen, *Monthly Weather Rev.* **87**, 319 (1959).
106. A. I. Wiin-Nielsen and I. A. Brown, *Proc. Int. Symp. Numerical Weather Prediction, Tokyo*, Meteorol. Soc. Japan, Tokyo, 1962, p. 593.
107. A. I. Wiin-Nielsen, I. A. Brown, and M. Drake, *Tellus* **15**, 261 (1963).
108. A. I. Wiin-Nielsen, I. A. Brown, and M. Drake, *Tellus* **16**, 168 (1964).
109. A. I. Wiin-Nielsen, *WMO-IUGG Symp. Res. and Develop. Aspects of Long-Range Forecasting*, WMO Tech. Note No. 66, 1965, p. 177.
110. J. Perloth, *Tellus* **14**, 403 (1962).

111. C. B. Pyke, *Bull. Am. Meteorol. Soc.* **46**, 4 (1965).
112. D. L. Bradbury, *J. Appl. Meteorol.* **1**, 421 (1962).
113. V. G. Semenov and G. M. Shushevskaya, *Tr. Tsentr. Inst. Progn.* **115**, 141 (1962).
114. V. G. Semenov, *Meteorol. i Gidrol.*, No. 4, 24 (1963).
115. R. Scherhag, *Geophysica (Helsinki)* **6**, No. 3–4, (1958).
116. R. Scherhag, in *Dinamika krupnomasshtabnykh atmosfernykh protsessov* [The dynamics of large-scale atmospheric processes], Nauka, 1967, p. 233.
117. J. Namias, *J. Geophys. Res.* **64**, 631 (1959).
118. J. Namias, *Proc. Int. Symp. Numerical Weather Prediction, Tokyo*, Meteorol. Soc. Japan, Tokyo, 1962, p. 615.
119. J. Namias, *J. Geophys. Res.* **68**, 6171 (1963).
120. J. Namias, *WMO-IUGG Symp. Res. and Develop. Aspects of Long-Range Forecasting*, WMO Tech. Note No. 66, 1965, p. 46.
121. J. S. Sawyer, ibid., p. 227.
122. E. G. Arkhipova, *Tr. Gos. Okeanogr. Inst.* [Trans. State Oceanogr. Inst.], No. 54, 35 (1960).
123. H. C. Schellard, Meteorol. Office London, Sci. Paper No. 11, 1962.
124. J. Bjerknes, *Geofys. Publikasjoner, Norske Videnskaps-Akad. Oslo* **24**, 115 (1962).
125. J. Bjerknes, *Proc. Rome Symp. Changes of Climate*, 1963, p. 297.
126. J. Bjerknes, *Proc. Symp. Arctic Heat Budget and Atmos. Circulation*, Mem. RM–5233–NSF, 1966, p. 473.
127. J. Bjerknes, in *Dinamika krupnomasshtabnykh atmosfernykh protsessov* [The dynamics of large-scale atmospheric processes] Nauka, 1967, p. 257.
128. J. Bjerknes, *WMO-IUGG Symp. Res. and Develop. Aspects of Long-Range Forecasting*, WMO Tech. Note No. 66, 1965, p. 77.
129. H. W. Ahlmann, Bowman Memorial Lecture I, Am. Geograph. Soc., New York, 1953.
130. H. H. Lamb and A. J. Johnson, *Geografiska Ann.*, 94 (1959); 363 (1961).
131. J. M. Mitchell, *Ann. N.Y. Acad. Sci.*, Art. 1, **95**, 235 (1961).
132. J. M. Mitchell, *Arid Zone Research* **20**, 161 (1963).
133. J. M. Mitchell, *Proc. Symp. Arctic Heat Budget and Atmos. Circulation*, Mem. RM–5233–NSF, 1966, p. 45.
134. V. N. Kolesnikova, and A. S. Monin, *Izv. Akad. Nauk SSSR, Fiz. Atmosfery i Okeana* **2**, 113 (1966).
135. A. M. Obukhov, "Weather and Turbulence," Presidential Address to JAMAP, XIV General Assembly of IUGG, Lucerne, Switzerland, 1967.
136. A. Defant, *Geografiska Ann.*, No. 3, (1921).
137. A. S. Monin, *Meteorol. i Gidrol.*, No. 7, 3 (1962).
138. E. N. Lorenz, *Proc. Int. Symp. Numerical Weather Prediction, Tokyo*, Meteorol. Soc. Japan, Tokyo, 1962, p. 629.
139. E. N. Lorenz, *Trans. N.Y. Acad. Sci.* **25**, 409 (1963).
140. E. N. Lorenz, *J. Atmos. Sci.* **20**, 130 (1963).

141. E. N. Lorenz, *Tellus* **16**, 1 (1964).
142. T. A. Sarymsakov, V. A. Dzhordzhio, and V. A. Bugayev, *Izv. Akad. Nauk SSSR, Ser. Geogr. i Geofiz.* **11**, 451 (1947).
143. A. S. Chaplygina, *Izv. Akad. Nauk SSSR, Ser. Geofiz.*, 1832 (1961).
144. P. D. Thompson, *Rept. XIth Gen. Assembly of the IUGG*, 1957.
145. E. A. Novikov, *Izv. Akad. Nauk SSSR, Ser. Geofiz.*, 1721 (1959).
146. L. A. Diky and T. D. Koronatova, *Meteorol. i Gidrol.* No. 5, 39 (1964).
147. P. D. Thompson, *Tellus* **9**, 69 (1957).
148. A. S. Monin, *Izv. Akad. Nauk SSSR, Ser. Geofiz.*, 1250 (1958).
149. A. M. Yaglom, in *Bernoulli–Bayes–Laplace Anniversary Volume*, ed. J. Neyman and L. M. LeCam, Springer, New York, 1965, p. 241.
150. H. Maruyama, *Papers Meteorol. Geophys. (Tokyo)* **12**, 216 (1961).
151. M. S. Eygenson, *Ocherki fiziko-geograficheskikh proyavleny solnechnoy aktivnosti* [Outlines of the physico-geographical manifestations of solar activity], Izd. Lvov Univ., 1957.
152. M. S. Eygenson, *Solntse, pogoda, i klimat* [Sun, weather, and climate], Gidrometizdat, 1963.
153. B. M. Rubashev, *Problemy solnechnoy aktivnosti* [Problems of solar activity], Nauka, 1964.
154. B. I. Sazonov, *Vysotnye baricheskiye obrazovaniya i solnechnaya aktivnost'* [High-altitude pressure formations and solar activity], Gidrometizdat, 1964.
155. R. Wolf, *Kortweg Sitzungsberichte*, Vienna, 1883.
156. H. E. Landsberg, J. M. Mitchell, and H. L. Crutcher, *Monthly Weather Rev.* **87**, 283 (1959).
157. G. W. Brier, *Ann. of N.Y. Acad. Sci.*, Art. 1, **95**, (1961).

Chapter 4

1. H. L. Kuo, *J. Meteorol.* **11**, 399 (1954); **13**, 82 and 521 (1956); **14**, 533 (1957).
2. D. Fultz, in *Compendium of Meteorology*, ed. T. F. Malone, Am. Meteorol. Soc., Boston, 1951, p. 1235.
3. D. Fultz, *J. Meteorol.* **8**, 263 (1951).
4. D. Fultz, *Advan. Geophys.* **7**, 1 (1961).
5. D. Fultz, R. R. Long, G. V. Owens, W. Bohan, R. Kaylor, and J. Weil, *Meteorol. Monographs* **4**, No. 21, (1959).
6. W. S. von Arx, *Progr. Phys. Chem. Earth* **2**, 1 (1957).
7. A. A. Dmitriyev, T. V. Bonchkovskaya, and N. L. Byzova, *Trudy konferentsii po modelirovaniyu protsessov v atmosfere i gidrosfere* [Proceedings conference on the modeling of processes in the atmosphere and hydrosphere], Izd. Acad. Nauk SSSR, 1962.
8. T. V. Bonchkovskaya, *Trudy vsesoyuznogo nauchno-meteorologicheskogo soveshchaniya*

[Proceedings all-union meteorological conference], vol. 2, Gidrometizdat, 1963, p. 153.

9. H. B. Squire, in *Surveys in Mechanics, G. I. Taylor Seventieth Anniversary Volume,* ed. G. K. Batchelor and R. M. Davies, Cambridge Univ. Press, New York, 1956, p. 139.

10. F. Vettin, *Ann. Physik* (2), 102 and 246 (1857).

11. R. Hide, *Quart. J. Roy. Meteorol. Soc.* **79**, 161 (1953).

12. H. Riehl and D. Fultz, ibid. **84**, 389 (1958).

13. W. W. Kellogg and C. Sagan, *The Atmospheres of Mars and Venus,* Publ. 944, Nat. Acad. Sci.—Nat. Res. Council, Washington, D.C., 1961.

14. C. Sagan and W. W. Kellogg, *Ann. Rev. Astronomy Astrophys.* **1**, 235 (1963).

15. G. P. Kuiper and B. M. Middlehurst, eds., *Planets and Satellites,* The Solar System, vol. 3, Univ. of Chicago Press, Chicago, 1961.

16. V. I. Moroz, *Fizika planet* [Physics of the planets], Nauka, 1967.

17. C. de Jager, *Encycl. Phys.* **52**, 80 (1959).

18. F. Ward, *Pure Appl. Geophys. (Milan)* **58**, 157 (1964).

19. L. A. Diky, *Izv. Akad. Nauk SSSR, Fiz. Atmosfery i Okeana* **1**, 469 (1965).

20. L. A. Diky, *Dokl. Akad. Nauk SSSR* **170**, 67 (1966).

21. G. S. Golitsyn and L. A. Diky, *Izv. Akad. Nauk SSSR, Fiz. Atmosfery i Okeana* **2**, 225 (1966).

22. G. S. Golitsyn and L. A. Diky, in *Dinamika krupnomasshtabnykh atmosfernykh protsessov* [The dynamics of large-scale atmospheric processes], Nauka, 1967, p. 200.

23. L. A. Diky and G. S. Golitsyn, *Tellus* **20**, 314 (1968).

24. S. L. Hess, *J. Meteorol.* **7**, No. 1, (1950).

25. F. A. Gifford, *Astrophys. J.* **123**, 159 (1956).

26. F. A. Gifford, *Monthly Weather Rev.* **92**, 435 (1964).

27. C. P. Prabhakara and I. S. Hogan, *J. Atmos. Sci.* **22**, 97 (1965).

28. C. B. Leovy and Y. Mintz, Mem. RM–5110–NASA, December 1966.

29. Y. Mintz, *WMO-IUGG Symp. Res. and Develop. Aspects of Long-Range Forecasting,* WMO Technical Note No. 66, 1965, p. 141.

30. Y. Mintz and A. Arakawa, *Trans. Am. Geophys. Union* **44**, 53 (1963).

31. Y. Mintz and A. Arakawa, *Proc. Symp. Arctic Heat Budget and Atmos. Circulation,* Mem. RM–5233–NSF, 1966, p. 369.

32. Y. Mintz, in *Dinamika krupnomasshtabnykh atmosfernykh protsessov* [The dynamics of large-scale atmospheric processes], Nauka, 1967, p. 139.

33. P. I. E. Peebles, *Astrophys. J.* **140**, 328 (1964).

34. R. Hide, *Nature* **190**, 895 (1961).

35. R. Hide, *Mem. Soc. Roy. Sci. Liège* **5**, 481 (1962).

36. R. Hide, *Planetary Space Sci.* **14**, 669 (1966).

37. R. Hide, in *Magnetism and the Cosmos,* ed. W. R. Hindmarsh et al., Am. Elsevier, New York, 1967, p. 378.

38. R. Hide, *Bull. Am. Meteorol. Soc.* **47**, 873 (1966).

39. F. Link, *Icarus* **6**, 129 (1967).
40. P. M. Stone, *J. Atmos. Sci.* **24**, 642 (1967).
41. A. S. Monin, *Izv. Akad. Nauk SSSR, Ser. Geofiz.*, 218 (1962).
42. A. S. Monin, *Meteorol. i Gidrol.*, No. 7, 3 (1962).
43. N. A. Bagrov, ibid., No. 2, 3 (1966).
44. G. V. Gruza, *Tr. Sredneaz. Nauchn-Issled. Gidrometinst.* [Trans. Central Asian Sci. Res. Hydrometeorol. Inst.] **29** (44), 3 (1967).
45. M. A. Omshansky, *Tr. Gl. Geofiz. Observ.* **14**, 49 (1937).
46. A. M. Obukhov, *Izv. Akad. Nauk SSSR, Ser. Geofiz.*, 339 (1955).

Index